D1239644

2018
THE STATE OF
FOOD SECURITY AND NUTRITION IN THE WORLD

BUILDING CLIMATE RESILIENCE FOR FOOD SECURITY AND NUTRITION

Food and Agriculture Organization of the United Nations

Rome, 2018

CONTENTS

TABLES, FIGURES AND BOXES

TABLES, FIGURES AND BOXES

FOREWORD

In September 2017, we jointly launched *The State of Food Security and Nutrition in the World*, marking the beginning of a new era in monitoring progress towards achieving a world without hunger and malnutrition, within the framework of the Sustainable Development Goals (SDGs).

This report monitors progress towards the targets of ending both hunger (SDG Target 2.1) and all forms of malnutrition (SDG Target 2.2), and provides an analysis of the underlying causes and drivers of observed trends. While *the prevalence of undernourishment* is at the forefront of monitoring hunger, *the prevalence of severe food insecurity* – based on the Food Insecurity Experience Scale (FIES) – was introduced last year to provide an estimate of the proportion of the population facing serious constraints on their ability to obtain safe, nutritious and sufficient food.

The report also tracks progress on a set of indicators used to monitor World Health Assembly global targets for nutrition and diet-related non-communicable diseases, three of which are also indicators of SDG2 targets.

The challenges we face are indeed significant. Of great concern is the finding last year that, after a prolonged decline, the most recent estimates showed global hunger had increased in 2016. Last year we observed that the failure to reduce world hunger is closely associated with the increase in conflict and violence in several parts of the world, and that efforts to fight hunger must go hand in hand with those to sustain peace. New evidence in this year's report corroborates the rise in world hunger, thus demanding an even greater call to action. Furthermore, while we must sow the seeds of peace in order to achieve food security, improve nutrition and "leave no one behind", we also need to redouble efforts to build climate resilience for food security and nutrition.

In 2017, the number of undernourished people is estimated to have reached 821 million – around one person out of every nine in the world. Undernourishment and severe food insecurity appear to be increasing in almost all subregions of Africa, as well as in South America, whereas the undernourishment situation is stable in most regions of Asia.

A more encouraging finding last year was that the rising trend in undernourishment had not yet been reflected in rates of child stunting; this continues to be the case this year. Nonetheless, we are concerned that in 2017, nearly 151 million children under five have stunted growth, while the lives of over 50 million children in the world continue to be threatened by wasting. Such children are at a higher risk of mortality and poor health, growth and development. A multisectoral approach is needed to reduce the burden of stunting and wasting, and to appropriately treat wasting to reduce childhood morbidity and mortality.

In addition to contributing to undernutrition, the food insecurity we are witnessing today also

FOREWORD

contributes to overweight and obesity, which partly explains the coexistence of these forms of malnutrition in many countries. In 2017, childhood overweight affected over 38 million children under five years of age, with Africa and Asia representing 25 percent and 46 percent of the global total, respectively. Anaemia in women and obesity in adults are also on the increase at the global level – one in three women of reproductive age is anaemic and more than one in eight adults – or more than 672 million – is obese. The problem of obesity is most significant in North America, but it is worrying that even Africa and Asia, which still show the lowest rates of obesity, are also experiencing an upward trend. Furthermore, overweight and obesity are increasing the risk of non-communicable diseases such as type 2 diabetes, high blood pressure, heart attacks and some forms of cancer.

In addition to conflict and violence in many parts of the world, the gains made in ending hunger and malnutrition are being eroded by climate variability and exposure to more complex, frequent and intense climate extremes, as shown in Part 2 of this report. Hunger is significantly worse in countries with agricultural systems that are highly sensitive to rainfall and temperature variability and severe drought, and where the livelihood of a high proportion of the population depends on agriculture. If we are to achieve a world without hunger and malnutrition in all its forms by 2030, it is imperative that we accelerate and scale up actions to strengthen the resilience and adaptive capacity of food systems and people's livelihoods in response to climate variability and extremes.

Building climate resilience will require climate change adaptation and disaster risk reduction and management to be integrated into short-, medium- and long-term policies, programmes and practices. National and local governments can find guidance in the outcomes and recommendations of existing global policy platforms: climate change (governed by the UNFCCC and the 2015 Paris Agreement); disaster risk reduction (the Sendai Framework on Disaster Risk Reduction); humanitarian emergency response (the 2016 World Humanitarian Summit and the Grand Bargain); improved nutrition and healthy diets (the Second International Conference on Nutrition [ICN2] and the UN Decade of Action on Nutrition 2016–2025); and development as part of the overarching 2030 Agenda for Sustainable Development. Currently many of these global policy platforms are still too compartmentalized and not well aligned. Therefore, we must do more to work towards a better integration of these platforms to ensure that actions across and within sectors such as environment, food, agriculture and health, pursue coherent objectives to address the negative impacts and threats that changing climate variability and increased climate extremes pose to people's food security, access to healthy diets, safe nutrition and health.

The transformative vision of the 2030 Agenda for Sustainable Development and the new challenges we face in ending hunger and malnutrition call on us to renew and strengthen our five organizations' strategic partnerships.

We reiterate our determination and commitment to step up concerted action to fulfil the ambitions of the 2030 Agenda and achieve a world free from hunger and all forms of malnutrition.

The alarming signs of increasing food insecurity and high levels of different forms of malnutrition are a clear warning that there is considerable work to be done to make sure we "leave no one behind" on the road towards achieving the SDG goals on food security and improved nutrition.

José Graziano da Silva
FAO Director-General

Gilbert F. Houngbo
IFAD President

Henrietta H. Fore
UNICEF Executive Director

David Beasley
WFP Executive Director

Tedros Adhanom Ghebreyesus
WHO Director-General

METHODOLOGY

The State of Food Security and Nutrition in the World 2018 has been prepared by the FAO Agricultural Development Economics Division in collaboration with the Statistics Division of the Economic and Social Development Department and a team of technical experts from FAO, IFAD, UNICEF, WFP and WHO.

A senior advisory team consisting of designated senior managers of the five UN publishing partners guided the production of the report. Led by FAO, this team decided on the outline of the report and defined its thematic focus. It further gave oversight to the technical writing team composed of experts from each of the five co-publishing agencies. The technical writing team involved external experts in preparing background papers to complement the research and data analysis undertaken by the members of the writing team.

The writing team produced a number of interim outputs, including an annotated outline, first draft and final draft of the report. These were reviewed, validated and cleared by the senior advisory team at each step in the preparation process. The final report underwent a rigorous technical review by senior management and technical experts from different divisions and departments within each of the five UN agencies, both at headquarters and decentralized offices. Finally, the report underwent executive review and clearance by the heads of agency of the five co-publishing partners.

ACKNOWLEDGEMENTS

The State of Food Security and Nutrition in the World 2018 was jointly prepared by the Food and Agriculture Organization of the United Nations (FAO), the International Fund for Agricultural Development (IFAD), the United Nations Children's Fund (UNICEF), the World Health Organization (WHO) and the World Food Programme (WFP).

Under the overall guidance of Kostas Stamoulis, the direction of the publication was carried out by Marco V. Sánchez Cantillo and José Rosero Moncayo, with the overall coordination of Cindy Holleman, the Editor of the publication, all of whom are from the FAO Economic and Social Development Department (ES), in collaboration with members of a Steering Committee formed by Paul Winters (IFAD), Victor Aguayo (UNICEF), Francesco Branca (WHO) and Arif Husain (WFP). Carlo Cafiero (FAO), Ama Brandford-Arthur and Ashwani Muthoo (IFAD), Chika Hayashi and Roland Kupka (UNICEF), Yvonne Forsén (WFP), and Marzella Wüstefeld (WHO) contributed to the coordination and provided technical editorial support. Valuable comments and final approval of the report were provided by the executive heads and senior staff of the five co-authoring agencies.

Part 1 of the report was coordinated by Anne Kepple (FAO). 1.1 was prepared by Carlo Cafiero with Klaus Grünberger, Anne Kepple and Sara Viviani (FAO). 1.2 was prepared by Chika Hayashi (UNICEF) and Laurence Grummer-Strawn (WHO) with Trudy Wijnhoven (FAO); Diana Estevez and Zita Weise Prinzo (WHO); and Julia Krasevec, Richard Kumapley, Vrinda Mehra, Louise Mwirigi and Nona Reuter (UNICEF). 1.3 was prepared by Anne Kepple (FAO) with Meghan Miller and Trudy Wijnhoven (FAO); Lena Hohfeld and Gaurav Singhal (WFP); Diana Estevez and Laurence Grummer-Strawn (WHO); and Chandana Maitra who prepared a background paper.

Part 2 of the report was coordinated by Cindy Holleman (FAO). 2.1 and 2.2 were prepared by Cindy Holleman with Trudy Wijnhoven (FAO); Tisorn Songsermsawas (IFAD); Lina Mahy and Marzella Wüstefeld (WHO); and inputs from a background paper prepared by Michele Meroni, Felix Rembold and Andrea Toreti (European Commission - Joint Research Centre [EC-JRC]) with Olivier Crespo, Bruce Hewitson, Christopher Jack, Pierre Kloppers and Mark Tadross (University of Cape Town [UCT]). 2.3 was prepared by Tania Osejo Carrillo (WFP) with Cindy Holleman (FAO); Giorgia Pergolini (WFP); and Lina Mahy, Zita Weise Prinzo and Marzella Wüstefeld (WHO). 2.4 was prepared by Sylvie Wabbes-Candotti, Julia Wolf (FAO) and Kathryn Millken (WFP) with Nora Boehm, Maryline Darmaun, Kaisa Kartunen, Rebeca Koloffon, Catherine Leclercq, Roman Malec and Trudy Wijnhoven (FAO); Cristina Colon, Cristina Klauth and Roland Kupka (UNICEF); Giorgia Pergolini (WFP); and Lina Mahy, Zita Weise Prinzo and Marzella Wüstefeld (WHO). Marco V. Sánchez Cantillo provided editorial inputs for Part 2.

Valuable comments and input on the report were provided by Lavinia Antonaci, Stephan Baas, Manuel Barange, Giacomo Branca, Jacqueline Demeranville, Dominique Burgeon, Rene Castro, Piero Conforti, Valentina Conti, Andre Croppenstedt, Juan Feng, Gustavo Gonzale, Elizabeth Graham, Uwe Grewer, Robert Guei, Adriana Ignaciuk, Anna Lartey, Sooyeon Jin, Alexander Jones, Panagiotis Karfakis, Markus Lipp, Giuseppe Maggio, Galimira Markova, Árni M. Mathiesen, Enrico Mazzoli, Meghan Miller, Jamie Morrison, Tamara Nanitashivili, Ana Ocampo, Oscar Rojas, Luca Russo, Antonio Scognamillo,

ACKNOWLEDGEMENTS

Salar Tayyib, Junko Sazaki, Barbara Sbrocca, David Sedik, Ramasamy Selvaraju, Josef Schmidhuber, Ahmed Shukri, Ilaria Sisto, Libor Stoukal, Andreas Thulstrup, Emilie Wieben, Natalia Winder-Rossi, Benoist Veillerette, Yahor Vetlou, Mario Zappacosta and Xia Zhang. (FAO); Frank Dentener and Mateo Zampieri (EC-JRC); Diane Holland and Dolores Rio (UNICEF); Constanza Di Nucci, Ilaria Firmian, Juliane Friedrich, Liza Leclerc, Nerina Muzurovic, Joyce Njoro, Lauren Philipps and Marian Odenigbo (IFAD); Rogerio Bonifacio and Katiuscia Fara (WFP); and Jonathan Abrahams, Adelheid Marschang, Karen McColl, Kim Petersen and Amy Savage (WHO).

Klaus Grünberger and Chiamaka Nwosu were responsible for preparing undernourishment and food security data, with input from Marinella Cirillo under the supervision of Carlo Cafiero and Sara Viviani. Supporting data were provided by Salar Tayyib and the Food Balance Sheets team of the FAO Statistics Division (ESS). Diana Estevez was responsible for consolidating the nutrition data, with input from Elaine Borghi, Laurence Grummer-Strawn, Lisa Rogers, Stefan Savin and Gretchen Stevens (WHO); and Richard Kumapley and Vrinda Mehra (UNICEF). Valentina Conti (FAO) was responsible for preparing the data analysis for Part 2 and Annex 2 and 3, with data inputs from Anne-Claire Thomas and Ferdinando Urbano (EC-JRC); conflict and food crisis data inputs from Aurelien Mellin (FAO); and climate data inputs from Khadra Ghedi Alasow, Luleka Dlamini, Fatima Mohamed, Kokesto Molepo and Tichaona Mukunga (UCT).

Support for report production came from Max Blanck, Andrew Park and Daniela Verona in the FAO Economic and Social Development Department.

The FAO Meeting Programming and Documentation Service provided printing services and carried out the translations.

The Publishing Group (OCCP) of the FAO Office for Corporate Communications provided editorial support, design and layout, as well as production coordination, for editions in all six official languages.

ACRONYMS AND ABBREVIATIONS

ASAP	Anomaly hot Spots of Agricultural Production
ASIS	Agricultural Stress Index System (FAO)
CCA	Climate change adaptation
CH	*Cadre Harmonisé* (harmonized framework)
CSA	Climate-smart agriculture
CV	Coefficient of variation
DEC	Dietary energy consumption
DES	Dietary energy supply
DRR	Disaster risk reduction
DRRM	Disaster risk reduction and management
EC-JRC	European Commission Joint Research Centre
ENSO	El Niño–Southern Oscillation
FAO	Food and Agriculture Organization of the United Nations
FIES	Food Insecurity Experience Scale
GAM	Global acute malnutrition
GHG	Greenhouse gas
GIEWS	Global Information and Early Warning System on Food and Agriculture
GIS	Geographic information systems
HLPE	High Level Panel of Experts on Food Security and Nutrition
ICN2	Second International Conference on Nutrition
IFAD	International Fund for Agricultural Development

IPC	Integrated Food Security Phase Classification
IPCC	Intergovernmental Panel on Climate Change
MDER	Minimum dietary energy requirement
MENA	Middle East and North Africa
NAP	National Adaptation Plan
NAPA	National Adaptation Programme of Action
NCD	Non-communicable disease
NDC	Nationally Determined Contribution
NDVI	Normalized difference vegetation index
PoU	Prevalence of undernourishment
SADC	Southern Africa Development Community
SD	Standard deviation
SDGs	Sustainable Development Goals
SFDRR	Sendai Framework for Disaster Risk Reduction
SIDS	Small Island Developing States
UCT	University of Cape Town
UNFCCC	United Nations Framework Convention on Climate Change
UNICEF	United Nations Children's Fund
USD	United States dollar
WFP	World Food Programme
WHA	World Health Assembly
WHO	World Health Organization

KEY MESSAGES

➜ New evidence continues to signal a rise in world hunger and a reversal of trends after a prolonged decline. In 2017 the number of undernourished people is estimated to have increased to 821 million – around one out of every nine people in the world.

➜ While some progress continues to be made in reducing child stunting, levels still remain unacceptably high. Nearly 151 million children under five – or over 22 percent – are affected by stunting in 2017.

➜ Wasting continues to affect over 50 million children under five in the world and these children are at increased risk of morbidity and mortality. Furthermore, over 38 million children under five are overweight.

➜ Adult obesity is worsening and more than one in eight adults in the world – or more than 672 million – is obese. Undernutrition and overweight and obesity coexist in many countries.

➜ Food insecurity contributes to overweight and obesity, as well as undernutrition, and high rates of these forms of malnutrition coexist in many countries. The higher cost of nutritious foods, the stress of living with food insecurity and physiological adaptations to food restriction help explain why food insecure families may have a higher risk of overweight and obesity.

➜ Poor access to food increases the risk of low birthweight and stunting in children, which are associated with higher risk of overweight and obesity later in life.

➜ Exposure to more complex, frequent and intense climate extremes is threatening to erode and reverse gains made in ending hunger and malnutrition.

➜ In addition to conflict, climate variability and extremes are among the key drivers behind the recent uptick in global hunger and one of the leading causes of severe food crises. The cumulative effect of changes in climate is undermining all dimensions of food security – food availability, access, utilization and stability.

➜ Nutrition is highly susceptible to changes in climate and bears a heavy burden as a result, as seen in the impaired nutrient quality and dietary diversity of foods produced and consumed, the impacts on water and sanitation, and the effects on patterns of health risks and disease, as well as changes in maternal care, child care and breastfeeding.

➜ Actions need to be accelerated and scaled up to strengthen resilience and adaptive capacity of food systems, people's livelihoods, and nutrition in response to climate variability and extremes.

➜ Solutions require increased partnerships and multi-year, large-scale funding of integrated disaster risk reduction and management and climate change adaptation programmes that are short-, medium- and long-term in scope.

➜ The signs of increasing food insecurity and high levels of different forms of malnutrition are a clear warning of the urgent need for considerable additional work to ensure we "leave no one behind" on the road towards achieving the SDG goals on food security and nutrition.

EXECUTIVE SUMMARY

ADVANCING FOOD SECURITY AND NUTRITION MONITORING IN THE ERA OF THE 2030 AGENDA FOR SUSTAINABLE DEVELOPMENT

Last year, *The State of Food Security and Nutrition in the World* marked the start of a new era in monitoring progress towards achieving a world without hunger and malnutrition in all its forms – an aim set out in the 2030 Agenda for Sustainable Development (2030 Agenda). Addressing the challenges of hunger, food insecurity and malnutrition in all its forms features prominently in the second Sustainable Development Goal (SDG) of the 2030 Agenda: Ensuring access to safe, nutritious and sufficient food for all (Target 2.1) and eliminating all forms of malnutrition (Target 2.2). It is also understood that attainment of SDG2 depends largely on – and also contributes to – the achievement of the other goals of the 2030 Agenda: ending poverty; improving health, education, gender equality and access to clean water and sanitation; decent work; reduced inequality; and peace and justice, to name only a few.

This transformational vision embedded in the 2030 Agenda provides an imperative for new ways of thinking, acting and measuring. For example, the growing global epidemic of obesity, which is increasingly affecting lower income countries and rapidly adding to the multiple burden of malnutrition and non-communicable diseases, also points to the need to re-examine how we think about and measure hunger and food insecurity as well as their linkages with nutrition and health. Fortunately, data gathering and measurement tools are rapidly evolving to meet the monitoring challenges presented by the new agenda.

Last year, this report included several innovations aimed at promoting new ways of thinking about food security and nutrition in the context of the 2030 Agenda and responding to the challenges of the Second International Conference on Nutrition (ICN2) Framework for Action and the UN Decade of Action on Nutrition 2016–2025. The scope of the report was expanded to include a set of six nutrition indicators used to monitor World Health Assembly global targets for nutrition and diet-related non-communicable diseases, three of which are also indicators of the SDG2 targets. The report also introduced for the first time a new indicator of food security, the prevalence of severe food insecurity based on the Food Insecurity Experience Scale (FIES), which is an estimate of the proportion of the population facing serious constraints on their ability to obtain sufficient food.

EVIDENCE CONTINUES TO POINT TO A RISE IN WORLD HUNGER IN RECENT YEARS, AN IMPORTANT WARNING THAT WE ARE NOT ON TRACK TO ERADICATE HUNGER BY 2030

Evidence continues to signal a rise in world hunger. According to available data, the number of people who suffer from hunger has been growing over the past three years, returning to levels from a decade ago. The absolute number of people in the world affected by undernourishment, or chronic food deprivation, is now estimated to have increased from around 804 million in 2016 to nearly 821 million in 2017. The situation is worsening in South America and most regions of Africa; likewise, the decreasing trend in undernourishment that characterized Asia until recently seems to be slowing down significantly. Without increased efforts, there is a risk of falling far short of achieving the SDG target of hunger eradication by 2030.

CHILD UNDERNUTRITION CONTINUES TO DECLINE, BUT LEVELS OF ADULT OBESITY AND ANAEMIA IN WOMEN OF REPRODUCTIVE AGE ARE INCREASING

Good nutrition is the lifeblood of sustainable development and drives the changes needed for a more sustainable and prosperous future.

Progress, although limited in magnitude and pace, has been made in reducing child stunting and increasing exclusive breastfeeding for the first six months of life. Nonetheless, while the prevalence of overweight in children under five years may not have changed significantly in recent years, adult obesity continues to rise and one in three women of reproductive age in the world is anaemic.

Children with low weight-for-height (wasting) have an increased risk of mortality. In 2017, 7.5 percent of children under five were affected by this form of undernutrition, with regional prevalences ranging from 1.3 percent in Latin America to 9.7 percent in Asia.

Multiple forms of malnutrition are evident in many countries. Poor access to food and particularly healthy food contributes to undernutrition as well as overweight and obesity. It increases the risk of low birthweight, childhood stunting and anaemia in women of reproductive age, and it is linked to overweight in school-age girls and obesity among women, particularly in upper-middle- and high-income countries. The higher cost of nutritious foods, the stress of living with food insecurity and physiological adaptations to food restriction help explain why food insecure families have a higher risk of overweight and obesity. Additionally, maternal and infant/child food deprivation can result in foetal and early childhood "metabolic imprinting", which increases the risk of obesity and diet-related non-communicable diseases later in life.

CLIMATE VARIABILITY AND EXPOSURE TO CLIMATE EXTREMES ARE THREATENING TO ERODE AND REVERSE GAINS MADE IN ENDING HUNGER AND MALNUTRITION

Having thoroughly investigated the role of conflict last year, the focus in 2018 is on the role of climate – more specifically, climate variability and extremes.

Climate variability and extremes are a key driver behind the recent rises in global hunger and one of the leading causes of severe food crises. The changing nature of climate variability and extremes is negatively affecting all dimensions of food security (food availability, access, utilization and stability), as well as reinforcing other underlying causes of malnutrition related to child care and feeding, health services and environmental health. The risk of food insecurity and malnutrition is greater nowadays because livelihoods and livelihood assets – especially of the poor – are more exposed and vulnerable to changing climate variability and extremes. What can be done to prevent this threat from eroding the gains made in ending hunger and malnutrition in recent years?

This report launches an urgent appeal to accelerate and scale up actions to strengthen resilience and adaptive capacity in the face of changing climate variability and increasing extremes. National and local governments are facing challenges in trying to determine measures to prevent risk and address the effects of these stressors. They can be guided by existing global policy platforms and processes whereby climate resilience is an important element: climate change (governed by the UNFCCC and the 2015 Paris Agreement); disaster risk reduction (the Sendai Framework on Disaster Risk Reduction); humanitarian emergency response (the 2016 World Humanitarian Summit and the Grand Bargain); improved nutrition and healthy diets (the Second International Conference on Nutrition [ICN2] and the UN Decade of Action on Nutrition 2016–2025); and development (as part of the overarching 2030 Agenda for Sustainable Development). However, it is important to ensure better integration of these global policy platforms and processes to ensure that actions across and within sectors such as environment, food, agriculture and health pursue coherent objectives. The success of policies, programmes and practices that national and local governments implement to address these challenges will also depend on cross-cutting factors, as well as specific tools and mechanisms that are adaptable to specific contexts.

Part 1 of this report presents the most recent trends in hunger, food insecurity and malnutrition in all its forms with a focus on monitoring progress on SDG Targets 2.1 and 2.2. This year the report also provides a deeper exploration of the indicator of wasting among children under five years of age. The last section of Part 1 aims to build the bridge between the first two sections by exploring the links between food insecurity and various forms of malnutrition. The current state of knowledge is presented on the pathways through which poor access to food can contribute simultaneously to undernutrition as well as overweight and obesity, resulting in the coexistence of multiple forms of malnutrition at the country level and even within the same households.

Part 2 closely scrutinizes the extent to which climate variability and extremes are undermining progress in the areas of food security and nutrition through different channels. The analysis ultimately points to guidance on how the key challenges brought about by climate variability and extremes can be overcome if we are to achieve the goals of ending hunger and malnutrition in all forms by 2030 (SDG Targets 2.1 and 2.2) as well as other related SDGs, including taking action to combat climate change and its impacts (SDG13).

TORIT, SOUTH SUDAN
Women from one of South Sudan's 60 agropastoral field school groups carry charcoal for cooking, part of an FAO-led project to improve nutrition and strengthen the resilience of households to food insecurity.
©FAO/Stefanie Glinski

PART 1
FOOD SECURITY AND NUTRITION AROUND THE WORLD IN 2018

FOOD SECURITY AND NUTRITION AROUND THE WORLD IN 2018

1.1 RECENT TRENDS IN HUNGER AND FOOD INSECURITY

KEY MESSAGES

➔ New evidence continues to point to a rise in world hunger in recent years after a prolonged decline. An estimated 821 million people – approximately one out of every nine people in the world – are undernourished.

➔ Undernourishment and severe food insecurity appear to be increasing in almost all regions of Africa, as well as in South America, whereas the undernourishment situation is stable in most regions of Asia.

➔ The signs of increasing hunger and food insecurity are a warning that there is considerable work to be done to make sure we "leave no one behind" on the road towards a world with zero hunger.

TARGET 2.1

"By 2030, end hunger and ensure access by all people, in particular the poor and people in vulnerable situations, including infants, to safe, nutritious and sufficient food all year round."

Prevalence of undernourishment

The 2017 edition of *The State of Food Security and Nutrition in the World* projected that the decade-long decline in the prevalence of undernourishment in the world had reached an end, and was possibly in reverse. This was largely attributed to persistent instability in conflict-ridden regions, adverse climate events that have hit many regions of the world and economic slowdowns that had affected more peaceful settings and worsened the food security situation. Now, new evidence confirms that lower levels of per capita food consumption in some countries, and increased inequality in the ability to access food in the populations of other countries, have contributed to what is projected to be **a further increase in the percentage of people in the world having insufficient dietary energy consumption in 2017**. The latest FAO estimates show that the share of undernourished people in the world population – the prevalence of undernourishment, or PoU – appears to have been growing for two years in a row, and may have reached 10.9 percent in 2017 (Figure 1 and Table 1).[1]

Even though the absolute increase in this percentage may seem negligible from a historical perspective, considering continuing population growth, it implies that the number of people who suffer from hunger has been growing over the past three years, returning to levels from almost a decade ago (Figure 1). The absolute number of undernourished people in the world is now estimated to have increased from around 804 million in 2016 to almost 821 million in 2017. This trend sends a clear warning that, if efforts are not enhanced, the SDG target of hunger eradication will not be achieved by 2030.

These new estimates (see Box 1) unfortunately confirm that the prevalence of undernourishment in Africa and Oceania has been increasing for a number of years (Table 1). Africa remains the continent with the highest PoU, affecting almost 21 percent of the population (more than 256 million people). The estimates also reveal

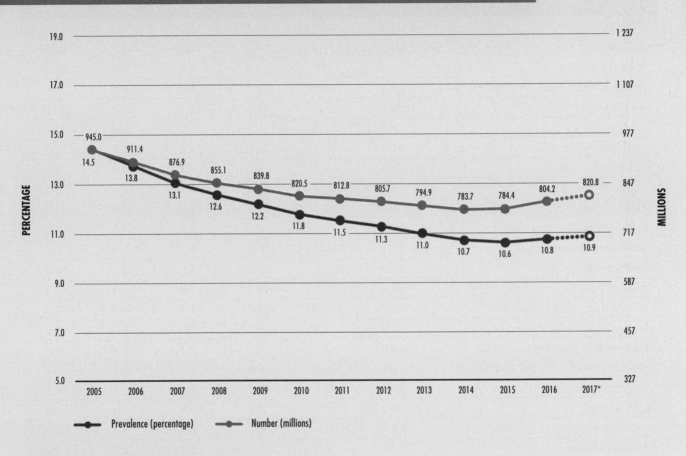

FIGURE 1
THE NUMBER OF UNDERNOURISHED PEOPLE IN THE WORLD HAS BEEN ON THE RISE SINCE 2014, REACHING AN ESTIMATED 821 MILLION IN 2017

Prevalence (percentage) — Number (millions)

* Projected values, illustrated by dotted lines and empty circles.
SOURCE: FAO.

that the decreasing trend that characterized Asia until recently may have come to an end. The projected PoU for Asia in 2017 points to a situation in which 11.4 percent of the population is estimated to be undernourished, which represents more than 515 million people, confirming it as the region with the highest number of undernourished people in the world.

A closer look at the subregions of Asia reveals that Western and South-eastern Asia are among those contributing to this slowdown in the decreasing trend, reflecting the fact that countries in South-eastern Asia have been affected by adverse climate conditions with

impacts on food availability and prices, while countries in Western Asia have been affected by prolonged armed conflicts.

In Africa, the situation is more pressing in the region of sub-Saharan Africa where an estimated 23.2 percent of the population – or between one out of four and one out of five people in the region – may have suffered from chronic food deprivation in 2017. An increase in the prevalence of undernourishment has been observed in all subregions of sub-Saharan Africa except for Eastern Africa. A further slight increase is seen in Southern Africa, while a significant uptick is seen in Western Africa,

TABLE 1
PREVALENCE OF UNDERNOURISHMENT IN THE WORLD, 2005–2017

	Prevalence of undernourishment (%)					
	2005	**2010**	**2012**	**2014**	**2016**	**2017[1]**
WORLD	14.5	11.8	11.3	10.7	10.8	10.9
AFRICA	21.2	19.1	18.6	18.3	19.7	20.4
Northern Africa	6.2	5.0	8.3	8.1	8.5	8.5
Northern Africa (excluding Sudan)	*6.2*	*5.0*	*4.8*	*4.6*	*5.0*	*5.0*
Sub-Saharan Africa	24.3	21.7	21.0	20.7	22.3	23.2
Eastern Africa	34.3	31.3	30.9	30.2	31.6	31.4
Middle Africa	32.4	27.8	26.0	24.2	25.7	26.1
Southern Africa	6.5	7.1	6.9	7.4	8.2	8.4
Western Africa	12.3	10.4	10.4	10.7	12.8	15.1
ASIA	17.3	13.6	12.9	12.0	11.5	11.4
Central Asia	11.1	7.3	6.2	5.9	6.0	6.2
South-eastern Asia	18.1	12.3	10.6	9.7	9.9	9.8
Southern Asia	21.5	17.2	17.1	16.1	15.1	14.8
Western Asia	9.4	8.6	9.5	10.4	11.1	11.3
Central Asia and Southern Asia	*21.1*	*16.8*	*16.7*	*15.7*	*14.7*	*14.5*
Eastern Asia and South-eastern Asia	*15.2*	*11.5*	*10.1*	*9.0*	*8.9*	*8.9*
Western Asia and Northern Africa	*8.0*	*7.1*	*8.9*	*9.3*	*9.9*	*10.0*
LATIN AMERICA AND THE CARIBBEAN	**9.1**	**6.8**	**6.4**	**6.2**	**6.1**	**6.1**
Caribbean	23.3	19.8	19.3	18.5	17.1	16.5
Latin America	8.1	5.9	5.4	5.3	5.3	5.4
Central America	8.4	7.2	7.2	6.8	6.3	6.2
South America	7.9	5.3	4.7	4.7	4.9	5.0
OCEANIA	5.5	5.2	5.4	5.9	6.6	7.0
NORTHERN AMERICA AND EUROPE	**< 2.5**	**< 2.5**	**< 2.5**	**< 2.5**	**< 2.5**	**< 2.5**

[1] Projected values.
SOURCE: FAO.

possibly reflecting factors such as droughts,[2] rising foods prices[3] and a slowdown of real per capita Gross Domestic Product (GDP) growth.[4] The dynamics in the prevalence of undernourishment, combined with rapid population growth, have led to a dramatic increase in the total number of undernourished people (Table 2). The number of undernourished people in sub-Saharan Africa rose from 181 million in 2010 to almost 222 million in 2016, an increase of 22.6 percent in six years, and – based on current projections – may have increased further to more than 236 million in 2017.

Although still in a context of a relatively low level of undernourishment, the situation is deteriorating in South America, where the PoU has increased from 4.7 percent in 2014 to a projected 5.0 percent in 2017. Such trends may be the result of persisting low prices in main export commodities – particularly crude oil – which have drained financial resources for food imports, reduced the capacity of governments to invest in the economy and significantly reduced the fiscal incomes needed to protect the most vulnerable against rising domestic prices and loss of income.

BOX 1
REVISED SERIES OF ESTIMATES OF THE PREVALENCE OF UNDERNOURISHMENT AND PROJECTIONS FOR 2017

In preparation for each edition of *The State of Food Security and Nutrition in the World*, the Statistics Division of FAO conducts a thorough revision of the entire series of PoU estimates, to reflect all updated or additional evidence gathered since the publication of the previous edition. As a result, the PoU series from different issues of the report cannot be directly compared; the reader is advised to refer to figures presented in the same issue to evaluate the evolution of undernourishment over time.

▶ In this edition, one major revision involves the **series of population data** used for all countries. National population figures are now obtained from the 2017 revision of the *World Population Prospects*[5] released by the United Nations Department of Economic and Social Affairs (DESA) Population Division in May 2017. It is worth noting that the new series of population estimates may present different figures also for earlier years, as official statistical series are revised retrospectively each time new data become available and inconsistencies are corrected. Population figures, both in terms of level and age/sex composition, have several implications for PoU estimates, as they enter into the computation of per capita levels of dietary energy supply (DES) and into estimates of the minimum dietary energy requirement (MDER) parameter and are used to calculate the number of undernourished people.
▶ This edition also includes updated DES estimates for a number of the countries with the largest undernourished populations in the world, resulting from a revision of the methodology used to compile the Food Balance Sheets.

As usual, PoU estimates are presented as three-year averages at the country level and as annual values at the regional and global level. Projections are needed in order to generate figures for the most recent time period. As in the past edition of *The State of Food Security and Nutrition in the World*, PoU estimates for 2017 are obtained by making a separate projection for each of the model's parameters: the dietary energy consumption (DEC), the coefficient of variation (CV) of this consumption and the minimum dietary energy requirement (MDER).

Projection of the DEC. The latest available data from national food balance sheets for most countries refer to a year between 2013 and 2016. To estimate a DEC value for the most recent years, data on the per capita availability of cereals and meats – available from the Trade and Market Division (EST) of FAO – are used to estimate the likely rates of change in per capita dietary energy availability from 2013, 2014, 2015 or 2016 (depending on the country) to 2017. Such rates of change are then applied to the latest available DEC values to project them up to 2017.

Projection of the CV. As no household survey data are available for 2017, in most countries the CV estimated from the last available food consumption survey data was projected up to 2017 with no change. However, when available, estimates of the prevalence of severe food insecurity – based on the Food Insecurity Experience Scale (FIES) – were used as auxiliary information in projecting the CV. Since 2014, FIES data provide timely evidence on changes in the prevalence of severe food insecurity (FI_{sev}) that might closely reflect changes in the PoU. Such changes can be used to make inferences regarding the likely changes in the CV that might have occurred in the most recent year. Detailed analysis by the FAO Statistics Division of PoU data and the underlying parameters shows that, on average, CVs explain about one-third of the differences in PoU after accounting for differences in DEC and MDER. Thus, for the countries that agreed to disseminate national estimates of their prevalence of food insecurity based on the FIES, changes in FI_{sev} from 2016 to 2017 were used to determine the likely changes in the CV over the same period. For those countries only, the CV was revised by the amount that would generate a change of 1 percent in the PoU every time a change of 3 percent is observed in FI_{sev}.

Projection of the MDER. The MDERs in 2017 are based on the projected population structures from *World Population Prospects* (2017 revision, medium variant) produced by the Population Division of UN DESA.

These projections are subject to revision in the future editions of this report as new data from surveys and new official data on Food Balance Sheets components become available. For further details, see the methodological note in Annex 1.

TABLE 2
NUMBER OF UNDERNOURISHED PEOPLE IN THE WORLD, 2005–2017

	Number of undernourished (millions)					
	2005	2010	2012	2014	2016	2017[1]
WORLD	**945.0**	**820.5**	**805.7**	**783.7**	**804.2**	**820.8**
AFRICA	**196.0**	**200.2**	**205.2**	**212.5**	**241.3**	**256.5**
Northern Africa	9.7	8.5	17.6	17.8	19.5	20.0
Sub-Saharan Africa	176.7	181.0	187.6	194.7	221.9	236.5
Eastern Africa	113.5	119.1	113.3	117.1	129.6	132.2
Middle Africa	36.2	36.5	36.4	36.1	40.8	42.7
Southern Africa	3.6	4.2	4.2	4.6	5.2	5.4
Western Africa	33.0	31.9	33.7	36.9	46.3	56.1
ASIA	**686.4**	**569.9**	**552.2**	**523.1**	**514.5**	**515.1**
Central Asia	6.5	4.6	4.0	4.0	4.2	4.4
Eastern Asia	219.1	178.4	160.4	142.6	139.5	139.6
South-eastern Asia	101.7	73.7	65.1	60.6	63.6	63.7
Southern Asia	339.8	293.1	299.6	289.4	278.1	277.2
Western Asia	19.4	20.1	23.1	26.5	29.1	30.2
Central Asia and Southern Asia	*346.3*	*297.7*	*303.7*	*293.4*	*282.3*	*281.6*
Eastern Asia and South-eastern Asia	*320.7*	*252.1*	*225.5*	*203.2*	*203.1*	*203.3*
Western Asia and Northern Africa	*29.1*	*28.6*	*40.7*	*44.3*	*48.6*	*50.1*
LATIN AMERICA AND THE CARIBBEAN	**51.1**	**40.7**	**38.9**	**38.5**	**38.9**	**39.3**
Caribbean	9.1	8.0	7.9	7.7	7.2	7.0
Latin America	42.1	32.6	31.0	30.8	31.7	32.3
Central America	12.4	11.6	11.9	11.6	11.0	11.0
South America	29.6	21.1	19.1	19.3	20.7	21.4
OCEANIA	**1.8**	**1.9**	**2.0**	**2.3**	**2.6**	**2.8**
NORTHERN AMERICA AND EUROPE[2]	**< 26.4**	**< 27.0**	**< 27.2**	**< 27.3**	**< 27.5**	**< 27.6**

[1] Projected values.
[2] Numbers for NORTHERN AMERICA AND EUROPE refer to less than 2.5 percent of the population each year.
SOURCE: FAO.

Prevalence of severe food insecurity in the population, based on the FIES

Last year, *The State of Food Security and Nutrition in the World* presented, for the first time, estimates of the prevalence of *severe* food insecurity based on the **Food Insecurity Experience Scale (FIES)**[6] (see Box 2).

The estimates are based on data collected by FAO using the FIES in more than 140 countries worldwide[7] and on data collected by national institutions using the FIES or other similar experience-based food security scales in a number of countries in the Americas, Africa and Asia.[8] National-level estimates have been calibrated against the **global FIES reference scale** to ensure worldwide comparability.[9] FIES results can be produced in a very timely manner, providing a real-time picture of the situation without being based on projections.

According to latest FAO estimates, in 2017, **close to 10 percent of the world population** »

BOX 2
HOW ARE HUNGER AND FOOD INSECURITY MEASURED?

Prevalence of undernourishment

The prevalence of undernourishment (PoU) is the traditional FAO indicator used to monitor hunger at the global and regional level. It was developed at a time when very few national governments, particularly in lower-income countries, collected data regularly on food consumption. The methodology relied on aggregated country-level data available for most countries and on the occasional data on food consumption available for a few countries, to produce an estimate of the proportion of the population that does not have regular access to enough dietary energy for a healthy, active life. Over time, thanks to progress in the implementation of national household surveys, the number of countries able to provide information on the inequality in access to food in their population has increased. Now more countries in the world collect information on people's access to food in periodic national population surveys, generating data that are increasingly being used to improve FAO country-level PoU estimates.

As most household surveys do not provide direct evidence on individual food consumption, the PoU is estimated using a statistical model where the distribution of habitual consumption is modelled for the population's representative individual. A caveat of the approach is that the inference can only be made at population group level, and disaggregated only to the point allowed by the representativeness of the surveys in which the data were collected. Given current data availability for most countries, PoU estimates cannot be produced at sufficiently disaggregated levels to be able to identify specific vulnerable populations within countries, which is a limitation for monitoring the very ambitious goal of zero hunger in an agenda that aims to "leave no one behind". Also, due to the probabilistic nature and the margins of uncertainty associated with the parameters of the model, which usually imply confidence intervals of about 5 percentage points around the estimate, the

PoU cannot monitor further progress in reducing hunger when levels of PoU are already very low.

The prevalence of severe food insecurity in the population based on the Food Insecurity Experience Scale

To complement the information provided by the PoU and to allow for monitoring SDG Target 2.1 globally in a more effective way, FAO took inspiration from countries already using a different approach to measuring food insecurity and scaled it up to the global level. The approach is based on asking people, directly in a survey, to report on the occurrence of conditions and behaviours that are known to reflect constraints on access to food. The Food Insecurity Experience Scale (FIES) survey module is composed of eight questions that have been carefully selected and tested, and proven effective in measuring the severity of the food insecurity situation of respondents in different cultural, linguistic and development contexts. FIES data are easy to process, so that results can be produced in a very timely manner, providing a real-time picture of the situation.

The FIES has two features that make it a valuable tool to meet the monitoring challenges presented by the 2030 Agenda. First, being a direct survey-based measure, when included in large-scale national population surveys, results can be disaggregated, thus helping identify which subpopulations within a country are most affected by food insecurity. Second, it is possible to estimate the prevalence of food insecurity at different levels of severity. Someone experiencing severe food insecurity is likely to have gone entire days without eating due to lack of money or other resources (see top figure next page).

Although based on different methods and sources of data, the PoU and the prevalence of severe food insecurity are both measures of the extent of severe food deprivation in the population (see Box 3 and Figure 4).

BOX 2
(CONTINUED)

FOOD INSECURITY BASED ON THE FIES: WHAT DOES IT MEAN?

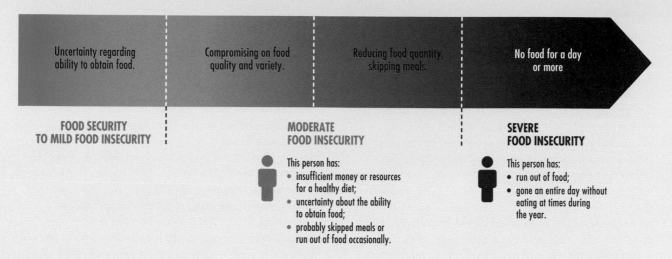

SOURCE: Created by FAO Statistics Division for this report.

FIGURE 2
SEVERE FOOD INSECURITY IS HIGHER IN 2017 THAN IT WAS IN 2014 IN EVERY REGION EXCEPT NORTHERN AMERICA AND EUROPE, WITH NOTABLE INCREASES IN AFRICA AND LATIN AMERICA

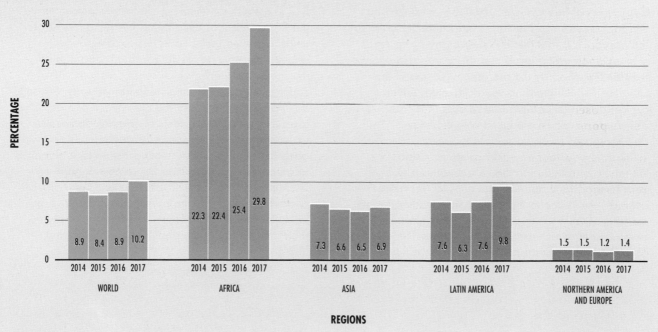

SOURCE: FAO.

TABLE 3
PREVALENCE OF SEVERE FOOD INSECURITY, MEASURED WITH THE FOOD INSECURITY EXPERIENCE SCALE, 2014–2017

	Prevalence (percentage in total population)			
	2014	**2015**	**2016**	**2017**
WORLD	**8.9**	**8.4**	**8.9**	**10.2**
AFRICA	**22.3**	**22.4**	**25.4**	**29.8**
Northern Africa	11.2	10.0	11.7	12.4
Sub-Saharan Africa	25.0	25.2	28.6	33.8
Eastern Africa	25.9	25.4	29.7	32.4
Middle Africa	33.9	34.3	35.6	48.5
Southern Africa	21.3	20.4	30.8	30.9
Western Africa	20.7	21.9	23.8	29.5
ASIA	**7.3**	**6.6**	**6.5**	**6.9**
Central Asia	1.9	1.7	2.7	3.5
Eastern Asia	< 0.5	< 0.5	0.9	1.0
South-eastern Asia	7.3	6.6	9.3	10.1
Southern Asia	13.5	12.0	10.1	10.7
Western Asia	8.8	9.0	9.4	10.5
Central Asia and Southern Asia	*13.0*	11.6	9.8	10.4
Eastern Asia and South-eastern Asia	2.4	2.2	3.3	3.6
Western Asia and Northern Africa	9.9	9.5	10.5	11.4
LATIN AMERICA AND THE CARIBBEAN	**n.a.**	**n.a.**	**n.a.**	**n.a.**
Caribbean	n.a.	n.a.	n.a.	n.a.
Latin America	7.6	6.3	7.6	9.8
Central America	12.7	10.2	8.3	12.5
South America	5.5	4.7	7.3	8.7
OCEANIA	**n.a.**	**n.a.**	**n.a.**	**n.a.**
NORTHERN AMERICA AND EUROPE	**1.5**	**1.5**	**1.2**	**1.4**

n.a. = data not available.
SOURCE: FAO.

» was exposed to **severe food insecurity, corresponding to about 770 million people**. At the regional level, values range from 1.4 percent in Northern America and Europe to almost 30 percent in Africa. As in the case of the PoU, severe food insecurity has been on the rise at the global level, driven by trends observed in Africa and Latin America (see Figure 2, Table 3 and Table 4).

It is important to note that the prevalence of severe food insecurity, based on the FIES, should not be confused with other indicators that use similar terminology to describe conditions of food insecurity (see Box 4).

Gender differences in food security

Examination of differences in development outcomes between men and women is particularly important to reveal where gender disparities exist, what their potential causes are, and how to address them. One interesting feature of data collected using the FIES module at the individual level is that it is possible to examine **gender differences in food security**.

Analysis of FIES data collected by FAO in more than 140 countries reveals that, in Africa, Asia and Latin America, the prevalence of severe food insecurity is slightly higher among women, with the largest differences found in Latin America (Figure 3). ∎

TABLE 4
NUMBER OF PEOPLE EXPERIENCING SEVERE FOOD INSECURITY, MEASURED WITH THE FOOD INSECURITY EXPERIENCE SCALE, 2014–2017

	Number (millions)			
	2014	2015	2016	2017
WORLD	**647.3**	**618.9**	**665.7**	**769.4**
AFRICA	**260.1**	**267.0**	**311.2**	**374.9**
Northern Africa	24.6	22.5	26.7	29.0
Sub-Saharan Africa	235.4	244.5	284.5	345.9
Eastern Africa	100.5	101.7	121.9	136.8
Middle Africa	50.6	52.7	56.5	79.2
Southern Africa	13.3	12.9	19.8	20.1
Western Africa	71.1	77.2	86.3	109.8
ASIA	**319.3**	**291.4**	**287.9**	**311.9**
Central Asia	1.3	1.1	1.9	2.5
Eastern Asia	<9.0	<9.1	15.3	16.4
South-eastern Asia	46.0	42.1	59.8	65.8
Southern Asia	242.2	218.1	186.2	199.2
Western Asia	22.3	23.2	24.7	28.0
Central Asia and Southern Asia	*243.5*	*219.3*	*188.1*	*201.7*
Eastern Asia and South-eastern Asia	*53.5*	*48.9*	*75.1*	*82.2*
Western Asia and Northern Africa	*46.9*	*45.7*	*51.5*	*57.0*
LATIN AMERICA AND THE CARIBBEAN	**n.a.**	**n.a.**	**n.a.**	**n.a.**
Caribbean	n.a.	n.a.	n.a.	n.a.
Central America	21.6	17.6	14.5	22.2
South America	22.8	19.4	30.8	36.7
NORTHERN AMERICA AND EUROPE	**16.2**	**16.3**	**13.5**	**15.2**

n.a. = data not available.
SOURCE: FAO.

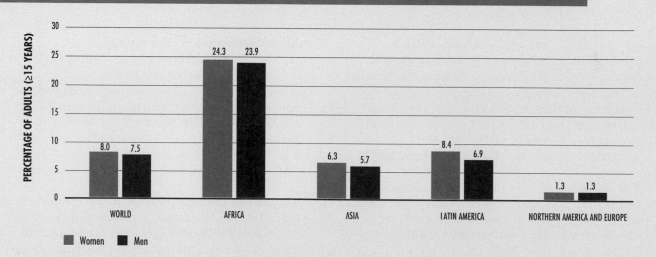

FIGURE 3
WOMEN ARE MORE LIKELY THAN MEN TO BE AFFECTED BY SEVERE FOOD INSECURITY IN AFRICA, ASIA AND LATIN AMERICA

PERCENTAGE OF ADULTS (≥15 YEARS)

	WORLD	AFRICA	ASIA	LATIN AMERICA	NORTHERN AMERICA AND EUROPE
Women	8.0	24.3	6.3	8.4	1.3
Men	7.5	23.9	5.7	6.9	1.3

■ Women ■ Men

SOURCE: FAO. 2018. Voices of the Hungry (2015–2017 three-year averages). In: *FAO* [online]. Rome. www.fao.org/in-action/voices-of-the-hungry

BOX 3
A COMBINED LOOK AT THE PREVALENCE OF UNDERNOURISHMENT AND OF SEVERE FOOD INSECURITY

Even though these two measures are based on different data and a different approach, the evidence provided by figures and trends in severe food insecurity, based on the FIES, is consistent with that provided by the series of figures on the PoU. This is not surprising when we consider that a condition of severe food insecurity, and the resulting reduction in the quantity of food consumed, might lead to the inability to cover dietary energy needs (i.e. the condition of "undernourishment" as defined in the PoU methodology). The combined analysis of the two indicators reinforces our belief that we are adequately capturing real trends.

Having two alternative views of the hunger problem also provides an important opportunity to cross-check the values of the two indicators for given countries. With reference to the average over the 2014–2016 period, the estimated prevalence of undernourishment and of severe food insecurity can be compared across a number of countries. The chart in Figure 4 shows that the two indicators provide a consistent picture for most countries, but still with some differences.

The chart is useful for identifying countries for which the difference between the two indicators is very large, pointing to the need for further investigation in order to detect potential data issues.[10] There are countries for which the PoU is much larger than the FI_{sev} (points in the lower-right section of the chart). In some cases, the estimated PoU may be too high because the CV could not be updated due to lack of access to recent survey data,[11] while in other cases, the FI_{sev} may be too low. In other countries (points in the upper-left section of the chart), the PoU may be underestimated or the FI_{sev} estimates may be too high.

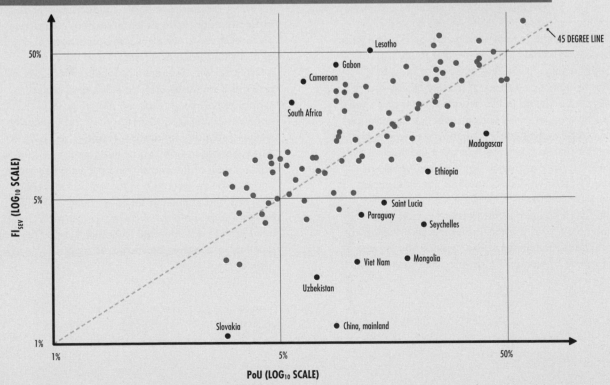

FIGURE 4
THE PREVALENCE OF UNDERNOURISHMENT AND THE PREVALENCE OF SEVERE FOOD INSECURITY SHOW A CONSISTENT PICTURE FOR MOST COUNTRIES, BUT DIFFERENCES EXIST

NOTES: The figure includes only countries for which the PoU is estimated to be larger than 2.5 percent and for which there is an estimate of the prevalence of severe food insecurity. A logarithmic scale of the data is used to highlight differences between smaller values.
SOURCE: FAO based on 2014–2016 three-year averages.

BOX 4
DIFFERENT FOOD SECURITY ASSESSMENTS FOR DIFFERENT OBJECTIVES

Since estimates of the prevalence of severe food insecurity (FI_{sev}) based on the FIES were first published in 2017, there is an ongoing need to clarify the relationship between this and other indicators that may use similar terminology to describe conditions of food insecurity. In particular, given the widespread use of the Integrated Food Security Phase Classification (IPC), it is common for people to request clarification regarding the relation between the number of people experiencing severe food insecurity estimated using the FIES and the number of people classified as acutely food insecure and in need of urgent action (Phase 3 or worse) in IPC reports (for examples of such reports, see www.ipcinfo.org).

The scope, methods, purpose and meaning of the numbers produced in the context of IPC analyses are different from the statistics produced for food security monitoring in the context of development agendas such as the SDGs. The most commonly known IPC scale is the Acute IPC analysis – this is the one referred to here. Percentages or numbers of acutely food-insecure people published in IPC reports cannot and should not be equated nor confused with the prevalence or numbers of severely food-insecure people based on the FIES (a component of SDG indicator 2.1.2, which is the prevalence of moderate or severe food insecurity). Understanding the differences between the two is critical for the correct utilization of each set of figures, as both are valuable for supporting strategic decision-making.

The SDG monitoring framework has the overall objective of monitoring development achievements and is based on reporting on a number of key, globally valid and comparable indicators. It relies on rigorous quantitative indicators, agreed upon by the Interagency and Expert Group on SDG indicators of the UN Statistical Commission. FIES data presented in this report are collected in the context of nationally representative surveys of the population, usually with a 12-month recall period. Measures obtained with the FIES are calibrated against a global reference scale of severity and used to estimate the prevalence of food insecurity in a globally comparable way.

IPC, on the other hand, has the specific objective of identifying populations in need of urgent action. To achieve this, IPC is based on convergence of evidence from a number of sources. To reach technical consensus on the classification of the severity of the food insecurity situation, a team of analysts conducts a critical evaluation and analysis of all available evidence on food security, which is compared against global standardized indicators and then interpreted with reference to local contexts. As a result, IPC analyses provide evidence needed to support emergency response planning. IPC analysis can be a snapshot of the food insecurity status in subnational areas – typically using data that is not older than two or three months – to give an overview of the current and projected situation and to provide information to decision-makers on ongoing and upcoming response needs. While extremely valuable for strategic response, IPC numbers are not intended to be used for monitoring achievements towards global development goals.

1.2 PROGRESS TOWARDS IMPROVING NUTRITION

KEY MESSAGES

→ Over 50 million children under five in the world are affected by wasting. Roughly half live in Southern Asia and one-quarter in sub-Saharan Africa. Addressing the burden of wasting will require a multipronged approach, including prevention, early identification, and treatment.

→ Progress has been made on reducing child stunting. However, nearly 151 million children under five in the world – or 22 percent – were still stunted in 2017, down from 25 percent in 2012, due mainly to progress in Asia. Over 38 million children under five are overweight.

→ Prevalences of anaemia in women and obesity in adults are increasing. More than one in eight adults in the world is obese and one in three women of reproductive age is anaemic.

TARGET 2.2

"By 2030, end all forms of malnutrition, including achieving, by 2025, the internationally agreed targets on stunting and wasting in children under five years of age, and address the nutritional needs of adolescent girls, pregnant and lactating women and older persons."

Nutrition is central to the 2030 Agenda. Target 2.2 calls for an end to all forms of malnutrition, and good nutrition also lays the foundation for achieving many of the SDGs (Figure 5). Improvements in nutrition directly support the achievement of ensuring healthy lives (SDG3), while also playing a role in ending poverty (SDG1), ensuring quality education (SDG4), achieving gender equality (SDG5), promoting economic growth (SDG8), and reducing inequalities (SDG 10). In this way, good nutrition is the lifeblood of

sustainable development, and drives the changes needed for a more sustainable and prosperous future.

In the 2012 World Health Assembly (WHA), Member States approved six global targets for improving maternal, infant and young child nutrition to be met by 2025. These WHA targets call for measures to: i) reduce anaemia in women of reproductive age; ii) reduce low birthweight in newborns; iii) increase rates of exclusive breastfeeding in infants; iv) reduce stunting; v) reduce wasting; and vi) halt the rise of overweight among children under five years of age. The latter three are also part of the SDG monitoring framework. To align with the 2030 deadline of the SDGs, this set of 2025 targets has been extended to 2030 to establish global targets for nutrition (see Box 5). In addition, the WHA plan of action for the prevention and control of non-communicable diseases also called for a reduction in adult obesity by 2025.

The State of Food Security and Nutrition in the World 2018 tracks progress on six of the seven above indicators. Low birthweight estimates will be released later in 2018 after the publication of this report and are thus not presented here.

Global trends

Globally, the proportion of children below the age of five who are stunted continues to decline, with 22.2 percent affected in 2017. The number of stunted children has also decreased from 165.2 million in 2012 to 150.8 million in 2017, representing a 9 percent decline over this five-year period. In 2017, 7.5 percent of children under five years of age – 50.5 million – suffered from wasting. Since 2012, the global proportion of overweight children seems stagnant, with 5.4 percent in 2012 (baseline year of WHA targets) and 5.6 percent (or 38.3 million) in 2017.

Globally, 36.9 percent of infants below six months of age were exclusively breastfed in 2012 (based on the most recent data for each country with data between 2005 and 2012), while 40.7 percent were exclusively breastfed in »

FIGURE 5
NUTRITION: ESSENTIAL TO ACHIEVE THE SUSTAINABLE DEVELOPMENT GOALS

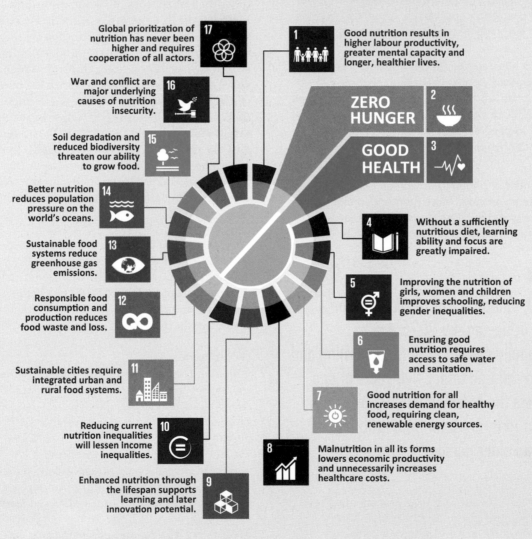

SOURCE: WHO Department of Nutrition for Health and Development, 2018.

BOX 5
EXTENDING THE WORLD HEALTH ASSEMBLY NUTRITION TARGETS TO 2030

In 2012, the World Health Assembly (WHA) agreed on six global targets for improving maternal, infant and young child nutrition to be achieved by 2025. Subsequently, in 2015, the Sustainable Development Goals established a global agenda for substantial improvement in nutrition by the year 2030, setting a specific objective of ending all forms of malnutrition by 2030, including achieving the 2025 targets and addressing the nutritional needs of adolescent girls, pregnant and lactating women, and older persons.

To align with the deadline year of 2030 for all SDG targets, UNICEF and WHO then extended the WHA nutrition targets up to the same year – in the process making some of them more ambitious – keeping in mind the original methodology used, the ambition declared in the SDGs to "end all forms of malnutrition", and the feasibility of achieving the new targets.[12]

The 2030 nutrition targets have been calculated based on a similar approach to that used for the 2025 targets. The rates of improvement between 1999 and 2017 were calculated for each indicator for all countries with trend data. After excluding countries that had already achieved a low level of malnutrition, the 20th percentile among all the rates of improvement was selected as an ambitious rate of improvement, but also one that has proven to be feasible in a large number of countries. This 20th percentile of the annual rate of improvement was then applied to the baseline prevalence globally to calculate a new 2030 target. Final numbers were rounded. For two of the indicators (low birthweight and anaemia in women of reproductive age), the past rate of improvement has been too slow to achieve the WHA target, even by 2030. Thus, for these indicators, the revised 2030 target is the same as the 2025 target, since the level of ambition for 2030 should not be less than that agreed upon for 2025.

For the other indicators, more ambitious targets for 2030 are proposed.

GLOBAL NUTRITION TARGETS REVISED FOR 2030 (FROM A 2012 BASELINE)

	2025 Target	2030 Target
Stunting	40% reduction in the number of children under five who are stunted.	50% reduction in the number of children under five who are stunted.
Anaemia	50% reduction in anaemia in women of reproductive age.	50% reduction in anaemia in women of reproductive age.
Low birthweight	30% reduction in low birthweight.	30% reduction in low birthweight.
Childhood overweight	No increase in childhood overweight.	Reduce and maintain childhood overweight to less than 3%.
Breastfeeding	Increase the rate of exclusive breastfeeding in the first six months up to at least 50%.	Increase the rate of exclusive breastfeeding in the first six months up to at least 70%.
Wasting	Reduce and maintain childhood wasting to less than 5%.	Reduce and maintain childhood wasting to less than 3%.

SOURCE: WHO and UNICEF. 2018. *The extension of the 2025 Maternal, Infant and Young Child nutrition targets to 2030*. Discussion paper.

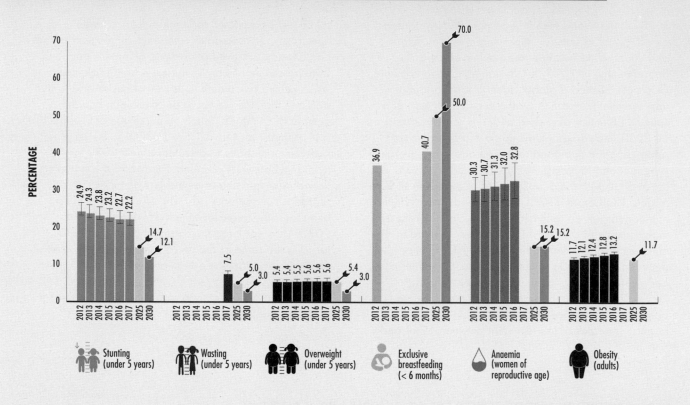

FIGURE 6

THERE IS STILL A LONG ROAD AHEAD TO ACHIEVE THE 2025 AND 2030 TARGETS FOR STUNTING, WASTING, OVERWEIGHT, EXCLUSIVE BREASTFEEDING, ANAEMIA IN WOMEN OF REPRODUCTIVE AGE AND ADULT OBESITY

SOURCES: Data for stunting, wasting and overweight are based on UNICEF, WHO and International Bank for Reconstruction and Development/World Bank. 2018. *UNICEF, WHO, World Bank Group Regional and Global Joint Malnutrition Estimates, May 2018 Edition* [online]. https://data.unicef.org/topic/nutrition, www.who.int/nutgrowthdb/estimates, https://data.worldbank.org; data for exclusive breastfeeding are based on UNICEF. 2018. Infant and Young Child Feeding: Exclusive breastfeeding, Predominant breastfeeding. In: *UNICEF Data: Monitoring the Situation of Children and Women* [online]. https://data.unicef.org/topic/nutrition/infant-and-young-child-feeding; data for anaemia are based on WHO. 2017. *Global Health Observatory (GHO)* [online]. http://apps.who.int/gho/data/node.imr.PREVANEMIA?lang=en; data for adult obesity are based on WHO. 2017. Global Health Observatory (GHO) [online]. http://apps.who.int/gho/data/node.main.A900A?lang=en

» 2017 (based on the most recent data for countries between 2013 and 2017).

It is shameful that one in three women of reproductive age globally is still affected by anaemia, with significant health and development consequences for both women and their children. The prevalence of anaemia among women of reproductive age has risen incrementally from 30.3 percent in 2012 to 32.8 percent in 2016. At the same time, adult obesity continues to rise each year, from

11.7 percent in 2012 to 13.2 percent in 2016, or 672.3 million people (Figure 6).

Regional patterns
Taking a closer look at the three SDG indicators, there are striking regional differences (Figure 7). While most regions seem to have made at least some progress towards the reduction of stunting prevalence between 2012 and 2017, Africa has seen the least progress in terms of relative improvement. In 2017, Africa and Asia accounted for more than nine out of ten of all »

FIGURE 7
DESPITE SOME PROGRESS TO REDUCE THE PREVALENCE OF STUNTED CHILDREN UNDER FIVE, MILLIONS ARE STILL AFFECTED BY STUNTING, WASTING AND OVERWEIGHT

NOTES: *Asia excluding Japan; **Oceania excluding Australia and New Zealand; ***The Global total factors in estimates for "more developed regions", but estimates for "more developed regions" are not displayed due to low population coverage.

SOURCE: UNICEF, WHO and International Bank for Reconstruction and Development/World Bank. 2018. *UNICEF, WHO, World Bank Group Regional and Global Joint Malnutrition Estimates, May 2018 Edition* [online]. https://data.unicef.org/topic/nutrition, www.who.int/nutgrowthdb/estimates, https://data.worldbank.org

BOX 6
LEVERAGING THE UNITED NATIONS DECADE OF ACTION ON NUTRITION 2016–2025

The United Nations Decade of Action on Nutrition 2016–2025, also referred to as the "Nutrition Decade", was declared by the UN General Assembly in 2016 to provide all stakeholders with a unique, time-bound opportunity to strengthen joint efforts and achieve a healthier and more sustainable future. Countries acknowledged the need for accelerated and sustained action to end malnutrition in all its forms, everywhere, leaving no one behind.

The first report on the implementation of the Nutrition Decade was presented by the UN Secretary-General to the UN General Assembly during its seventy-second session.[13] The report provides a review of the progress made in the implementation of national nutrition commitments. Currently, 183 countries have national policies that include nutrition goals and actions: 105 countries have health sector plans with nutrition components, 48 countries have national development plans with integrated nutrition objectives and about 70 countries have made efforts to mainstream food security and nutrition in sectoral policies and investment programmes. Moreover, 57 countries have implemented prevention and reduction of food insecurity risks, while 28 countries

have applied socio-economic measures to reduce vulnerability and strengthen resilience of communities at risk of climate hazards and emergencies.

However, in order to meet the global targets set, country implementation has to be scaled up, investments for nutrition need to be increased and enhanced policy coherence is required. The Nutrition Decade encourages governments to set country-specific SMART (specific, measurable, achievable, realistic, and time-bound) commitments for urgent investment, action and collaboration at national level. The first UN Secretary-General report calls for more actors and networks to join and engage, specifically city networks, communities acting on women's and children's health, human rights, water and climate change.[14]

The Nutrition Decade also provides a clearly defined, time-bound cohesive framework and is a space for aligned action on nutrition by all relevant actors. The Nutrition Decade provides countries with mechanisms such as Action Networks for sharing good practices, illustrating successes and challenges, promoting improved coordination and building political momentum to scale up global action.

More information about the Decade of Action on Nutrition can be found on www.un.org/nutrition

» stunted children globally, representing 39 percent and 55 percent respectively. Africa has seen an upward trend in the number of stunted children, while Asia has experienced the largest relative decrease in stunting prevalence. The confidence limits around the estimates for Oceania are too large to make clear conclusions.

In 2017, 50.5 million children under five were affected by wasting, with two regions – Asia and Oceania – seeing almost one in ten affected, compared to just one in one hundred in Latin America and the Caribbean. Most of the burden is concentrated in Asia, with seven out of ten wasted children in the world residing in that region.

In 2017, childhood overweight affected 38.3 million children, with Africa and Asia

representing 25 percent and 46 percent of the global total respectively, despite being the regions with the lowest percentage of children who are overweight (5.0 percent in the Africa region and 4.8 percent in Asia). Oceania (8.7 percent) and Latin America and the Caribbean (7.3 percent) have the highest prevalence. There has not been a significant change in overweight prevalence or numbers affected for any region between 2012 and 2017.

Rates of exclusive breastfeeding in Africa and Asia are 1.5 times those in Northern America where only 26.4 percent of infants under six months receive breastmilk exclusively. Conversely, the prevalence of anaemia among women of reproductive age in Africa and Asia is nearly three times higher than in Northern America. No region has shown a decline in anaemia among women of reproductive age. Adult obesity is highest in Northern America and the rate of increase in adult obesity is also the highest there. While Africa and Asia continue to have the lowest rates of obesity, there, too, they are increasing (see Annex 1).

General conclusions

Overall, there has been some progress regarding stunting and exclusive breastfeeding, although it may not be sufficient to achieve the global nutrition targets. Conversely, the scenarios for childhood overweight, adult obesity and anaemia among women of reproductive age are not improving.

To achieve the WHA 2025 and SDG 2030 nutrition targets will require increased investment in nutrition interventions, scaled-up implementation of policies and programmes, enhanced policy coherence, and a greater number of national commitments.

Global attention on addressing malnutrition in all its forms is unprecedented, with ICN2 galvanizing nations around a clear action agenda. As a follow-up to ICN2, the UN Decade of Action on Nutrition 2016–2025 has become an umbrella framework for countries to share experiences, promote improved coordination and build political momentum to scale up action towards eliminating malnutrition in all its forms (see Box 6). The Scaling Up Nutrition movement, comprising 60 countries, continues to galvanize multisectoral action to end stunting and all forms of malnutrition. All partners and stakeholders can coordinate efforts around this momentum to scale up nutrition interventions and work towards eliminating malnutrition.

Spotlight on wasting

Wasting is defined as having a low weight-for-height ratio according to the WHO Child Growth Standards.[15] Specifically, wasting is defined as weight-for-height below minus two standard deviations, and severe wasting is defined as weight-for-height below minus three standard deviations, from the median weight-for-height in the reference population. Wasting reflects a reduction or loss of body weight and is considered a relevant indicator of acute malnutrition. Additional indicators of acute

malnutrition are small mid-upper arm circumference and bilateral pitting oedema. This year's *The State of Food Security and Nutrition in the World* report takes a closer look at the problem of wasting among children under five years of age.

Global targets for wasting are to reduce the prevalence below 5 percent by 2025 and below 3 percent by 2030. In 2017, 7.5 percent of children under five were affected by wasting, with regional prevalences ranging from 1.3 percent (Latin America and the Caribbean) to 9.7 percent (Asia). Across all regions, around one-third of all children identified as wasted were severely wasted, with the exception of Latin America and the Caribbean, where one-quarter of those affected suffered from severe wasting (Figure 8 and Figure 9).

Children affected by wasting have a higher risk of mortality. An analysis from 2013 indicated that 875 000 deaths (or 12.6 percent of all deaths) among children under five years of age were related to wasting, of which 516 000 deaths (7.4 percent of all deaths among under-fives) were related to severe wasting.[16] Whereas the mortality risk associated with wasting is highest in the first few years of life, low weight-for-height continues to be a nutritional problem even for older children (see Box 7).

The main underlying causes of wasting are poor household food security, inadequate feeding and care practices, and/or poor access to health, water, hygiene and sanitation services. Suboptimal breastfeeding, poor complementary foods and poor feeding practices can lead to rapid weight loss or growth failure. Lack of knowledge about proper food storage, preparation and consumption by parents and caregivers may be contributing factors. Wasting may be part of a vicious cycle with infection: undernutrition increases the susceptibility to infection, and infection then leads to greater weight loss due to appetite loss and poor intestinal absorption. Diarrhoeal disease, in particular, often leads to rapid weight loss, and poor access to appropriate and timely health care slows the recovery from such illnesses. It is not yet well understood how much wasting contributes to conditions such as stunting, low

FIGURE 8
RATES OF CHILD WASTING REMAIN EXTREMELY HIGH IN SOME SUBREGIONS IN 2017, ESPECIALLY IN ASIA

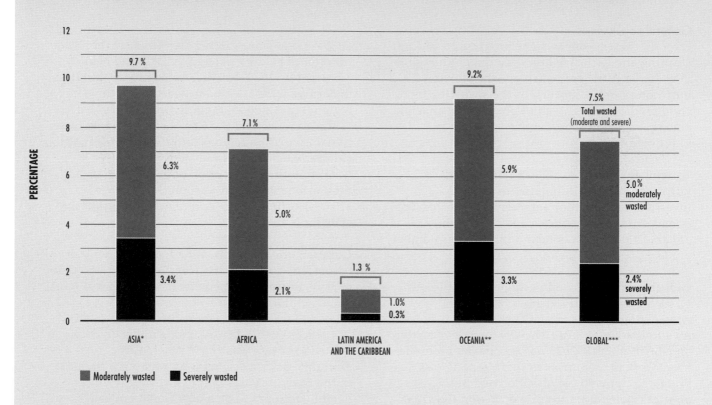

NOTES: *Asia excluding Japan; **Oceania excluding Australia and New Zealand; ***The Global total factors in estimates for "more developed regions", but estimates for "more developed regions" are not displayed due to low population coverage. Differences in total are due to rounding of figures to the nearest decimal point.
SOURCE: UNICEF, WHO and International Bank for Reconstruction and Development/The World Bank. 2018. *Levels and trends in child malnutrition: key findings of the 2018 Edition of the Joint Child Malnutrition Estimates* [online]. https://data.unicef.org/topic/nutrition, www.who.int/nutgrowthdb/estimates, https://data.worldbank.org

birthweight and anaemia. Evidence does suggest,[17] however, that episodes of wasting negatively affect linear growth and, therefore, undermine child growth and development.

All of the underlying causes of undernutrition described above can be exacerbated in humanitarian crisis situations, as they often have a negative impact on the quantity and diversity of foods available to children and women. This is particularly dangerous in resource-poor settings where ongoing food scarcity leads to monotonous child diets with low nutrient density that constrain

child growth. Furthermore, humanitarian crisis situations often restrict access to health care, and water and sanitation facilities, leading to a concomitant increase in diseases.

Wasting is typically measured in terms of its prevalence at the time of a survey. However, because wasting is often a short-term condition compared to other forms of malnutrition, the prevalence at a point in time underestimates the number of new cases that occur during an entire calendar year (i.e. incidence). Estimates of wasting prevalence can vary across seasons.

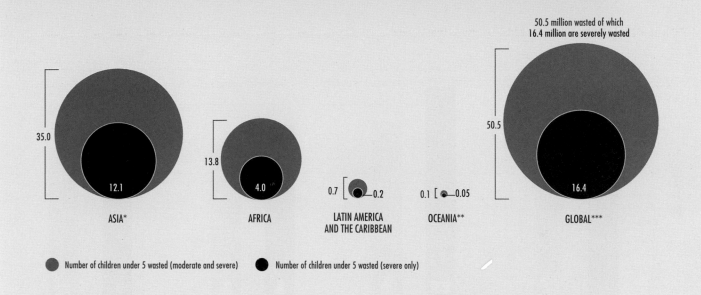

FIGURE 9
MILLIONS OF CHILDREN ARE AT INCREASED RISK OF MORTALITY DUE TO WASTING IN 2017, MAINLY IN ASIA AND AFRICA

50.5 million wasted of which
16.4 million are severely wasted

35.0

13.8

50.5

12.1

4.0

0.7 [●—0.2

0.1 [●—0.05

16.4

ASIA*

AFRICA

LATIN AMERICA
AND THE CARIBBEAN

OCEANIA**

GLOBAL***

● Number of children under 5 wasted (moderate and severe)   ● Number of children under 5 wasted (severe only)

NOTES: *Asia excluding Japan; **Oceania excluding Australia and New Zealand; ***The Global total factors in estimates for "more developed regions", but estimates for "more developed regions" are not displayed due to low population coverage.
SOURCE: UNICEF, WHO and International Bank for Reconstruction and Development/The World Bank. 2018. *Levels and trends in child malnutrition: key findings of the 2018 Edition of the Joint Child Malnutrition Estimates* [online]. https://data.unicef.org/topic/nutrition, www.who.int/nutgrowthdb/estimates, https://data.worldbank.org

They are commonly at their highest during the rainy season, often coinciding with the preharvest period and thus with food scarcity as well as higher rates of diseases including diarrhoea and malaria. Hazard events including protracted and acute emergencies can also affect wasting rates – therefore context needs to be considered in addition to seasonality. Documenting trends in wasting prevalence over time is difficult, as surveys are not generally conducted at the same time of year within all regions of a given country.

It is estimated that 50.5 million children globally under five suffer from wasting at any given point in time. Roughly half of these live in Southern Asia and an additional one-quarter in sub-Saharan Africa. Countries with a prevalence above 15 percent (very high category)[18] include Djibouti, Eritrea, India, the Niger, Papua New Guinea, South Sudan, Sri Lanka, the Sudan, and Yemen. While wasting is often thought of as a problem in emergency situations, the majority of children affected by wasting live outside of the emergency context.

Wide variations in prevalence of wasting exist between countries but also within countries, where wasting rates are on average 1.4 times higher among children from the poorest households. Aggregate figures do not indicate notable differences in the prevalence of wasting between girls and boys under five or by their place of residence or maternal education (Figure 10), although significant differences have been reported in specific countries and settings.

Disparities in the prevalence of child wasting between the richest and poorest households are

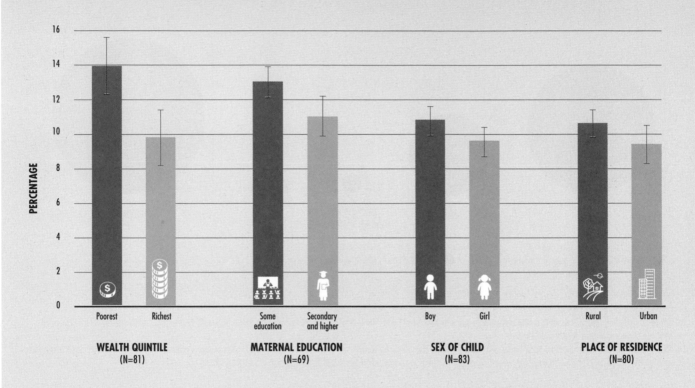

FIGURE 10
INEQUALITIES IN INCOME, EDUCATION, GENDER AND PLACE OF RESIDENCE REFLECT ON CHILD WASTING RATES

NOTE: Estimates are based on a subset of countries with disaggregated data between 2012 and 2018; each pair of demographic characteristics is based on a different subset of countries (N = number of countries).
SOURCE: UNICEF, WHO and World Bank. 2018. *Joint Child Malnutrition expanded country database, May 2018* [online].
https://data.unicef.org/topic/nutrition/malnutrition/

observed in many subregions (Figure 11). In three of five subregions in Africa, the poorest have significantly higher rates of wasting – nearly twice as high in Eastern Africa – when compared to the richest. In subregions with lower rates such as Central America and Southern Africa, there is no notable difference between the richest and the poorest.

Within countries, large differences in wasting prevalence can be observed between geographical regions. Figure 12 shows regions with the highest and lowest prevalence of wasting in a selection of countries where wasting prevalence is 10 percent or above at the national level. In some surveys,

such as those in the Gambia and Yemen, there is no significant difference in wasting prevalence between the geographical regions with the highest and lowest prevalence. In others, such as Chad, Nigeria and the Sudan, large differences exist. However, the prevalence of wasting may not be entirely comparable among geographic regions given that estimates may be based on data collected in different seasons owing to differences in survey timing and duration across different regions for any given country.

The potential effect of seasonal variation on under-five wasting rates can be particularly important in countries like India, where data

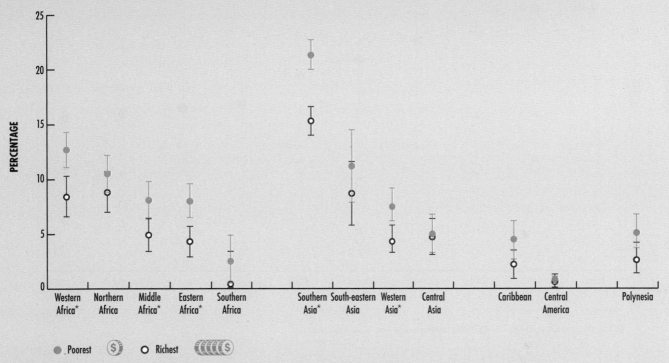

FIGURE 11
DISPARITIES IN THE PREVALENCE OF CHILD WASTING ARE OBSERVED BETWEEN THE POOREST AND THE RICHEST HOUSEHOLDS, ESPECIALLY IN EASTERN AFRICA

NOTES: Estimates are based on countries with disaggregated data between 2012 and 2018. Only regions with sufficient population coverage are displayed.
*Differences between the prevalence of wasting among the poorest and the richest quintiles are statistically significant.
SOURCE: UNICEF, WHO and World Bank. 2018. *Joint Child Malnutrition expanded country database, May 2018* [online].
https://data.unicef.org/topic/nutrition/malnutrition/

collection for the National Family and Health Survey 2015–16 (NFHS 2015–16) spanned an entire year. During a full year, India experiences several seasonal variations – such as harvest season, droughts and rains – which may affect wasting prevalence. Thus, the large geographical difference in the prevalence of child wasting observed in India may be influenced by when the survey was conducted in specific regions. However, other factors may also contribute to the wide gaps noted in prevalence of wasting by state. For instance, in the state with the highest prevalence of child wasting, about 70 percent of the households do not have access to sanitation facilities and almost half (46.1 percent) of the population belongs to India's poorest wealth quintile. In contrast, in the state with the lowest

prevalence of child wasting, nearly all households (99 percent) have access to sanitation facilities, though a majority (63.7 percent) of that state's population belongs to India's richer wealth quintiles.

Because wasting is often inaccurately considered to be a condition that occurs only during emergency situations, ongoing programmes to address this form of malnutrition outside of the emergency context are typically inadequate in scale and often in quality. In 2016, over 4 million children under the age of five years were admitted to treatment programmes for severe wasting – a large increase since 2014, when just over 3 million were admitted.[19] However, with an estimated 17 million severely wasted children at

FIGURE 12
LARGE DIFFERENCES EXIST IN PREVALENCE OF CHILD WASTING WITHIN REGIONS AND COUNTRIES

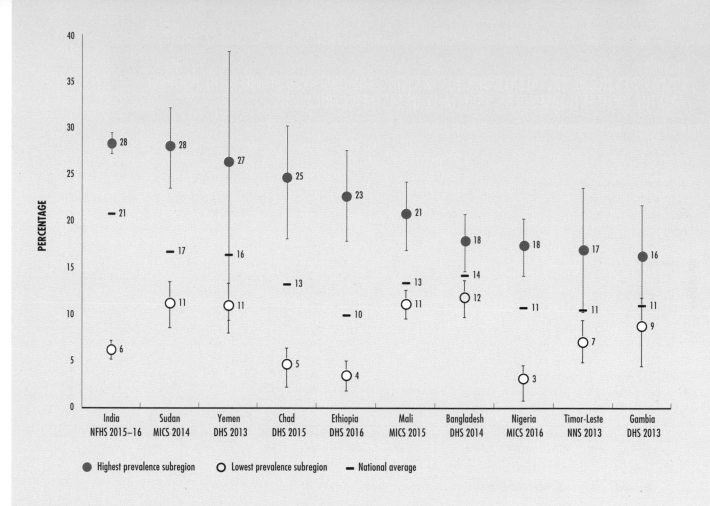

● Highest prevalence subregion   ○ Lowest prevalence subregion   ▬ National average

SOURCE: UNICEF, WHO and World Bank. 2018. *Joint Child Malnutrition expanded country database* [online]. https://data.worldbank.org, https://data.unicef.org/topic/nutrition/malnutrition/#access_data, http://apps.who.int/nutgrowthdb/database/search/Dataset/Search

any given point in time during the year 2016, far too few (i.e. one in four) were admitted into these life-saving programmes. Funding to care for children with severe wasting is often short-term and is focused primarily on humanitarian situations. Sustainable and adequately resourced programmes to prevent malnutrition in all its forms are necessary to reach the SDG targets for nutrition, including child wasting.

Addressing the burden of wasting will require a multipronged approach, including prevention in infancy and early childhood, early identification before children develop medical complications, and treatment of affected children, particularly those with severe wasting. An analysis from 2013 indicated that management of acute malnutrition, combined with the delivery of an infant and young child nutrition package – including the protection, promotion and support of appropriate breastfeeding, good complementary foods and feeding practices, and micronutrient supplements – scaled up to 90 percent coverage, could reduce the prevalence of severe wasting by 61.4 percent.[20]

Prevention of wasting requires addressing the underlying causes of malnutrition. Breastfeeding support and nutrition counselling for families – particularly regarding how to improve the quality of complementary foods and feeding practices – and early care for common childhood illnesses are essential. Food systems need to ensure that they deliver nutritious, safe and affordable diets for infants and young children, »

BOX 7
THINNESS AMONG SCHOOL-AGE CHILDREN

Whereas the mortality risk associated with wasting is highest in the first few years of life, low weight-for-height continues to be a nutritional problem even for older children. Thinness among children 5–9 years of age and adolescents 10–19 years of age is associated with higher risk of infectious diseases, delayed maturation, and reduced muscular strength, work capacity and bone density later in life.[21] Improved nutrition leads to better growth, development and educational achievements in school-age children.[22] For girls, thinness (defined as low Body Mass Index (BMI) for age) is associated with adverse pregnancy outcomes including maternal mortality, delivery complications, preterm birth, and intrauterine growth retardation.

Globally, over 10 percent of children aged 5–19 have a BMI-for-age below -2 standard deviations from the median of the WHO reference population. As is the case with wasting among preschool-age children, there are dramatic differences in the prevalence of thinness among children aged 5–19 years by region of the world. Thinness in school-age children is extremely high in India, where over one-quarter of children are too thin. The prevalence is

also high (>15 percent) in Afghanistan, Bangladesh, Bhutan, Nepal, Pakistan and Sri Lanka. The global prevalence of thinness has barely changed over the past decade, with less than a percentage point drop in prevalence since 2005.

School nutrition programmes can be an effective platform for providing nutritious meals or snacks, micronutrient supplements, and nutrition information, education and counselling. There is also a need for more nutrition intervention programmes among school-age children in addition to programmes for preschool children. Schools are increasingly being recognized as an effective platform for providing nutrition and health interventions to school-age children and adolescents. School feeding programmes can help prevent hunger, increase school enrolment, reduce absenteeism and improve learning outcomes. Interventions such as deworming and micronutrient supplementation are also linked to better nutrition and learning. The promotion of good nutrition and health in school settings is viewed as an effective tool to improve the growth and development of children and reduce risk factors for non-communicable diseases. In addition, SDG2 highlights the importance of nutrition for adolescent girls.

PREVALENCE OF THINNESS AMONG SCHOOL-AGE CHILDREN – 2016

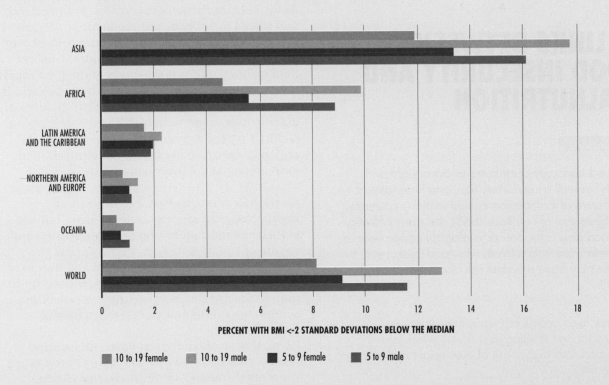

PERCENT WITH BMI <-2 STANDARD DEVIATIONS BELOW THE MEDIAN

■ 10 to 19 female ■ 10 to 19 male ■ 5 to 9 female ■ 5 to 9 male

Source: WHO. 2018. Global Health Observatory (GHO) data. In: *World Health Organization* [online]. www.who.int/gho

» including the most vulnerable. Water, hygiene and sanitation programmes need to ensure access to safe drinking water and sanitation facilities. Furthermore, social protection and safety net programmes need to ensure access to healthy diets for children and families left behind by mainstream development.

Improved growth monitoring and promotion, for instance by vaccination services and during child health and nutrition days, could help identify children at risk of severe wasting and at risk of morbidity and mortality, such as those who are moderately wasted and those in need of treatment. Furthermore, medical and nutritional treatment of severe acute malnutrition needs to be scaled up as part of routine health and nutrition services for children to improve childhood survival. These key areas are outlined in the sets of recommended actions in the ICN2 Framework for Action that countries are encouraged to implement, as appropriate, under the umbrella of the UN Decade of Action on Nutrition. ■

1.3 LINKS BETWEEN FOOD INSECURITY AND MALNUTRITION

KEY MESSAGES

➔ Food insecurity contributes to overweight and obesity, as well as undernutrition, and high rates of these forms of malnutrition coexist in many countries. The higher cost of nutritious foods, the stress of living with food insecurity, and physiological adaptations to food restriction help explain why food-insecure families may have a higher risk of overweight and obesity.

➔ Poor food access increases the risk of low birthweight and stunting in children, which are associated with higher risk of overweight and obesity later in life.

➔ Access to safe, nutritious and sufficient food must be framed as a human right, with priority given to the most vulnerable. Policies that promote nutrition-sensitive agriculture and food systems are needed, with special attention to the food security and nutrition of children under five, school-age children, adolescent girls and women in order to halt the intergenerational cycle of malnutrition.

At first glance, the preceding sections may appear to be telling different stories, confirming the trends described in the 2017 *State of Food Security and Nutrition in the World* report: hunger and food insecurity are on the rise, while child stunting continues to decline. In addition, the prevalence of obesity among adults in the world increased steadily between 1975 and 2016 – and at an accelerated pace over the past decade. How can these seemingly contradictory food security and nutrition trends be reconciled?

The focus on child wasting in the preceding section shows the challenges involved in building knowledge on the relationship between food insecurity and nutrition outcomes. Like child wasting, the causes of child stunting and other forms of malnutrition are complex, multisectoral and rooted in political and economic structures and ideological factors that influence control over resources.[23] When looked at through the lens of food systems, additional aspects of the food supply chain, food environment and consumer behaviour that influence the basic and underlying causes of malnutrition come to light.[24] These interplaying factors vary from context to context – across regions, countries, areas within countries, and even among and within households.

As the discussion in Part 2 of this report highlights, food security is a necessary but not sufficient condition to prevent malnutrition and ensure adequate nutrition. Part 2 depicts the complex interplay of multiple food and non-food factors affecting nutritional status, including the four dimensions of food security – availability, access, utilization and stability (see Figure 28).

Using this as an analytical basis, this section takes a closer look at one small part of the larger conceptual framework of causes and effects of food security and nutrition: the pathways from food access to malnutrition. This is important because different pathways can lead to nutritional outcomes as disparate as stunting in children and obesity in adults. Such details are seldom captured in existing conceptual figures linking food security and nutrition, yet they are essential for illuminating the mechanisms by which food insecurity can lead to different manifestations of malnutrition. Awareness of these pathways is crucial for understanding observed trends and designing effective policies and programmes aimed at improving nutrition.

Following is an overview of the current body of knowledge on the relationship between food insecurity (specifically the experience of not having *access* to safe, nutritious and sufficient food due to lack of money or other resources) and selected indicators of malnutrition. The different pathways from food insecurity to malnutrition are discussed in detail to dispel misunderstandings about the apparent hunger-obesity paradox and to shed light on the implications for policy-making. The aim is to advance the discussion of food security and nutrition to align it with the ambitions of the 2030 Sustainable Development Agenda.

The nutrition transition, food insecurity and the multiple burden of malnutrition

The trends presented in the previous section are characteristic of the global nutrition transition.[25] Rapid demographic, social and economic changes in many low- and middle-income countries have led to increased urbanization and changes in food systems, lifestyles and eating habits. As a consequence, dietary patterns have shifted toward increased consumption of processed foods that are often energy-dense, high in saturated fats, sugars and salt, and low in fibre.

Such changes bring with them a shift in the profile of nutritional status and diet-related diseases. In pre-transition conditions, the nutritional problems that predominate among the more vulnerable population groups are undernutrition and nutrient deficiencies. The transition gradually brings about increased energy consumption in the population, including among the more vulnerable. Under such conditions, undernutrition and some nutrient deficiencies begin to decline, while the excessive consumption of energy-dense, processed foods high in fats, salt and sugars becomes a major issue. These consumption habits lead to increasing rates of overweight and diet-related non-communicable chronic diseases, such as cardiovascular disease and diabetes. Table 5 shows how dietary and nutritional profiles change over three stages of the nutrition transition.

In this context, while large inequalities in the levels of childhood stunting and wasting persist across regions and countries, a simultaneous increase in overweight and obesity is observed, often in the same countries and communities with relatively high levels of child stunting. This coexistence of undernutrition with overweight and obesity is commonly referred to as the "double burden" of malnutrition.[26] Moreover, overweight and obese individuals can also be affected by micronutrient (vitamin and mineral) deficiencies, often called "hidden hunger"

TABLE 5
STAGES OF THE NUTRITION TRANSITION

Characteristic	Stages		
	Pre-transition	Transition	Post-transition
Diet (prevalent)	Grains, tubers, vegetables, fruits	Increased consumption of sugar, fats and processed foods	Processed foods with high content of fat and sugar; low fibre content
Nutritional problems	Undernutrition and nutritional deficiencies predominate	Undernutrition, nutritional deficiencies and obesity coexist	Overweight, obesity and hyperlipidaemia predominate

SOURCE: Adapted from C. Albala, S. Olivares, J. Salinas and F. Vio. 2004. *Bases, prioridades y desafíos de la promoción de salud.* Santiago, Universidad de Chile, Instituto de Nutrición y Tecnología de los Alimentos. [Bases, priorities and challenges of promoting health. Santiago, University of Chile, Institute of Nutrition and Food Technology].

FIGURE 13
COUNTRIES AFFECTED BY MULTIPLE FORMS OF MALNUTRITION

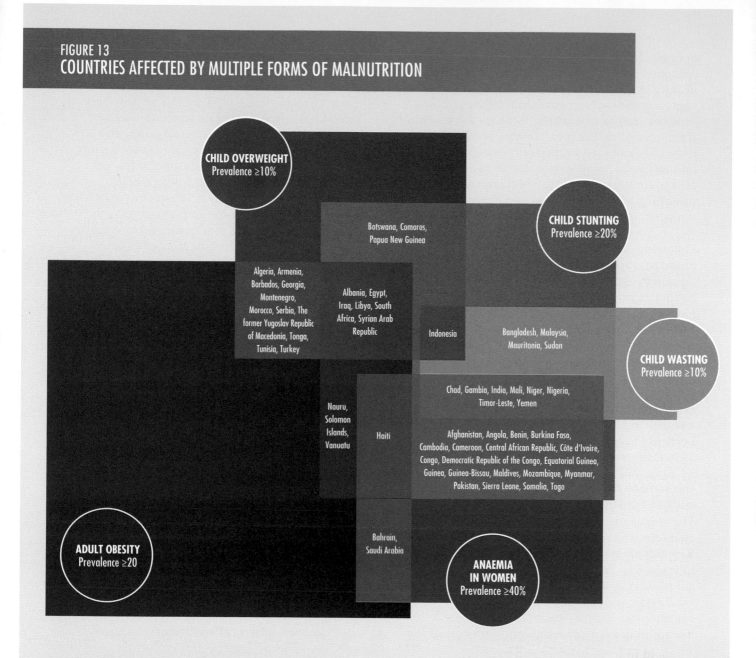

CHILD OVERWEIGHT
Prevalence ≥10%

CHILD STUNTING
Prevalence ≥20%

Botswana, Comoros,
Papua New Guinea

Algeria, Armenia,
Barbados, Georgia,
Montenegro,
Morocco, Serbia, The
former Yugoslav Republic
of Macedonia, Tonga,
Tunisia, Turkey

Albania, Egypt,
Iraq, Libya, South
Africa, Syrian Arab
Republic

Indonesia

Bangladesh, Malaysia,
Mauritania, Sudan

CHILD WASTING
Prevalence ≥10%

Chad, Gambia, India, Mali, Niger, Nigeria,
Timor-Leste, Yemen

Nauru,
Solomon
Islands,
Vanuatu

Haiti

Afghanistan, Angola, Benin, Burkina Faso,
Cambodia, Cameroon, Central African Republic, Côte d'Ivoire,
Congo, Democratic Republic of the Congo, Equatorial Guinea,
Guinea, Guinea-Bissau, Maldives, Mozambique, Myanmar,
Pakistan, Sierra Leone, Somalia, Togo

Bahrain,
Saudi Arabia

ADULT OBESITY
Prevalence ≥20

**ANAEMIA
IN WOMEN**
Prevalence ≥40%

NOTE: Only countries with at least one data point on nationally representative data since 2005 were included in the figure. There is low population coverage of high-income countries as only 14 countries had data on stunting, 3 on wasting and 15 on overweight in children under five years of age. The figure includes only names of countries with high prevalence of more than one form of malnutrition. The size of each box is proportionate to the total number of countries with a high prevalence of the respective form of malnutrition.
SOURCE: Created by FAO and WHO based on the most updated country data available from UNICEF, WHO and World Bank. 2018. *Joint child malnutrition estimates – Levels and trends (2018 edition)* [online]. www.who.int/nutgrowthdb/estimates for wasting, stunting and overweight in children under five years of age; for anaemia, WHO. 2017. *Global Health Observatory (GHO)* [online]. http://apps.who.int/gho/data/node.imr.PREVANEMIA?lang=en, and database on anaemia in: *World Health Organization* [online]. www.who.int/vmnis/database/anaemia; for adult obesity, WHO. 2017. *Global Health Observatory (GHO)* [online]. http://apps.who.int/gho/data/node.main.A900A?lang=en

because there may be no visible signs. It is estimated that 1.5 billion people in the world are affected by one or more forms of micronutrient deficiency.[27] Iron deficiency

anaemia in women of reproductive age is a form of micronutrient deficiency that can be present even in women who are overweight or appear to be well nourished.

Figure 13 shows countries that have a high prevalence of more than one form of malnutrition. The size of each box is proportionate to the total number of countries with a high prevalence of the respective form of malnutrition: child stunting, 73 countries; child wasting, 14; child overweight, 29; adult obesity, 101; and anaemia in women of reproductive age, 35. The prevalence threshold that is considered high for child stunting is 20 percent or above; for child wasting and child overweight, the threshold is 10 percent or higher.[28] Among these countries, Indonesia is the only one that shows a high prevalence of all three of these forms of child malnutrition, whereas nine countries have simultaneously high prevalence of child stunting and child overweight. Six of these nine countries also have a prevalence of adult obesity exceeding 20 percent, which is considered to be a high threshold. Eleven countries have simultaneously high rates of overweight among children and prevalence of adult obesity above 20 percent.

With respect to anaemia among women of reproductive age, WHO considers a prevalence of 40 percent or higher to be of severe public health significance.[29] Three countries have a high prevalence of anaemia in women and levels of adult obesity above 20 percent, and one of these – Haiti – also suffers from a high prevalence of child stunting. Twenty-nine countries have a high prevalence of anaemia in women and also of child stunting, with eight of these also suffering from a high prevalence of child wasting.

The multiple burden of malnutrition is more prevalent in low-, lower-middle and middle-income countries and concentrated among the poor. Obesity in high-income countries is similarly concentrated among the poor.[30] The coexistence of multiple forms of malnutrition can occur not only within countries and communities, but also within households – and can even affect the same person over their lifetime. Various examples of such situations are found at the household and individual level. A household may, for instance, have both a stunted child and an overweight or obese mother. At an individual level, a woman could be both overweight and suffer from anaemia, and a child could be simultaneously stunted and overweight.[31]

Food insecurity, in terms of poor food access, contributes to these situations in ways that are not always obvious. Moderate levels of food insecurity are often associated with diets that are energy-dense yet poor in micronutrients, because resource constraints may force people to reduce the nutritional quality of their diets. For example, these diets can cause micronutrient deficiencies in children that impede their growth and may also lead to obesity in mothers. At the same time, a diet that increases obesity may be lacking in iron, and thus may result in both obesity and anaemia in the same woman.

Pathways from food insecurity to malnutrition

There are multiple pathways whereby the experience of food insecurity – defined here as uncertain access to sufficient, safe and nutritious food – may contribute to forms of malnutrition as seemingly divergent as undernutrition and obesity. Figure 14 illustrates details of the link between food access and nutritional outcomes that are difficult to capture in comprehensive conceptual frameworks depicting the many basic, underlying and immediate causes of food insecurity and malnutrition.

As the figure shows, the main pathways from food insecurity to malnutrition pass through food consumption, or diet. Indicators of dietary intake are crucial to understanding the pathway from food insecurity to nutritional outcomes. More information on the food environment and food intake is needed to shed light on this relationship.

Figure 14 illustrates a number of key links and nexuses that make up the pathways from food insecurity to malnutrition. Two pathways are depicted: one leading from food insecurity to undernutrition and another leading to overweight and obesity. Below they are examined in more detail along with the evidence from studies that investigated these links using experience-based food insecurity metrics in combination with indicators of nutritional status.[32]

The food insecurity–undernutrition link. This link – from poor food access to child stunting and wasting and micronutrient deficiencies – is more easily understood,

FIGURE 14
PATHWAYS FROM INADEQUATE FOOD ACCESS TO MULTIPLE FORMS OF MALNUTRITION

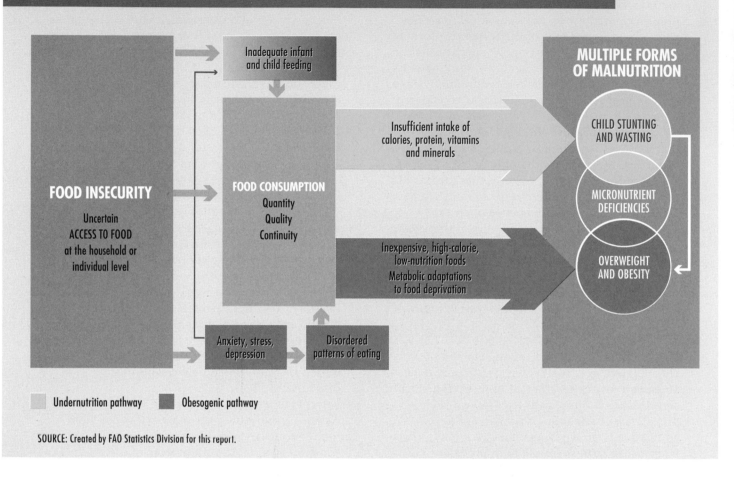

SOURCE: Created by FAO Statistics Division for this report.

because it is intuitive. A diet characterized by insufficient intake of calories, protein, vitamins and minerals will impede foetal, infant and child growth and development. Such diets contribute to maternal undernutrition and consequently to higher risk of low birthweight, which in turn are both risk factors for child stunting.

Existing research points to a link between household food insecurity and stunting among children (Table 6).[33] A majority of 30 studies reviewed examining this relationship found that food insecurity was strongly associated with negative effects on child linear growth in Africa, Asia and Latin America, whereas a few studies in Northern America found no association.[34] Although most studies clearly show a link, the

association between food insecurity and stunted growth may be obscured in the recent global trends in stunting, because the latter are based on stunting data available for many countries that were collected many years prior to the FIES data. Unless actions are taken immediately, signs of the recent increase in severe food insecurity may become evident in the regional and global trends in stunting in the near future.

One factor that increases a child´s risk of becoming stunted is low birthweight. Household food insecurity has been found to be associated with low birthweight in infants, in low-income as well as high-income settings.[35] However, the number of studies that have examined the link between food insecurity and low birthweight is still limited.

TABLE 6
SUMMARY OF FINDINGS OF STUDIES INCLUDED IN A LITERATURE REVIEW OF THE LINKS BETWEEN EXPERIENCED FOOD INSECURITY AND SELECTED FORMS OF MALNUTRITION

Association between food insecurity and (nutrition indicator)	Number of studies	Number of studies and association found		Differences in results by income level of country
		Association found	No association found	
Child wasting	15	3 positive 1 mixed*	11	No apparent difference.
Child stunting				
<5 years of age	21	16 positive 1 mixed*	4	Relatively more lower-middle and upper-middle-income countries report significant association compared to lower income countries.
≥5 years of age/ school-age	9	4 positive 2 mixed*	3	Studies showing no association are mostly from upper-middle- and high-income countries.
Child overweight				
<5 years of age	13	2 positive 2 mixed*	9	Association limited or absent in low- and lower-middle-income countries.
≥5 years of age/ school-age	21	3 positive 7 mixed* 1 negative	10	
Low birthweight	3	3 positive	0	No apparent difference.
Adult obesity				
Studies of women only	15	7 positive	8	Positive association predominant in high-income countries.
Studies that included both men and women	8	5 positive (in women only)	3	
Anaemia in women of reproductive age	8	6 positive 1 mixed*	1	No apparent difference.

NOTE: *Mixed means positive association in some groups only.
SOURCE: C. Maitra. 2018. *A review of studies that examine the link between food insecurity and malnutrition.* Technical Paper. Rome, FAO.

Little evidence is currently available supporting the association between food insecurity and child wasting. Three out of fifteen studies on this link reported a positive association, mostly in low- and lower-middle-income countries (Table 6).[36] As discussed in the preceding section, wasting is an indicator of acute malnutrition, which is strongly influenced by factors other than food insecurity (such as infections and diseases often caused by a lack of access to safe water, sanitation and quality health services). Child wasting may also be the result of short-term shocks and humanitarian crises.

Food insecurity is a risk factor for anaemia in women of reproductive age. Six out of eight studies reviewed from diverse countries and regions found a significant association between the two (Table 6).[37]

The stress of living with food insecurity can also have a negative effect on the nutrition of infants by compromising breastfeeding. Exclusive breastfeeding in the first six months protects against child stunting and wasting as well as against obesity later in life.[38] The existing evidence suggests that infants in food-insecure households are at a higher risk of not being exclusively breastfed.[39] Household food insecurity is associated with higher rates of maternal depression and stress in lower-middle-income as well as high-income countries, and this can undermine maternal confidence and self-efficacy, adversely affecting initiation and duration of breastfeeding and age-appropriate complementary feeding.[40]

Thus, as shown in Figure 14, food insecurity can both directly (through compromised diets) and indirectly (through the impact of stress on infant feeding) cause child wasting, stunting and micronutrient deficiencies. Nutritional knowledge and food habits may play a role by moderating the effects of household food insecurity on diet and, consequently, on nutritional outcomes. Finally, it is important to bear in mind that lack of access to clean water, sanitation and quality health care can cause diarrhoea and infectious diseases that interfere with the body´s ability to absorb nutrients. Recurrent infections and disease are serious contributing factors to wasting and stunting in children.

The food insecurity–obesity link. Although it may appear to be a paradox, food insecurity is often associated with overweight and obesity. As such, it may lead policy-makers in countries where many of the poor and food insecure are overweight to question the allocation of resources for food assistance. However, the association between food insecurity and overweight and obesity is in fact not contradictory, and can be understood by considering the obesogenic pathway presented in Figure 14.

The link between food insecurity and overweight and obesity passes through diet, which is affected by the cost of food. Nutritious, fresh foods often tend to be expensive. Thus, when household resources for food become scarce, people choose less expensive foods that are often high in caloric density and low in nutrients, particularly in urban settings and upper-middle- and high-income countries. In the context of globalized food markets where the relative cost of foods that are high in fats and sugar is low compared to fresh products such as fruits, vegetables and legumes, the prioritization of cost

for food-insecure families may result in diets high in energy and low in diversity, micronutrients and fibre.[41] Food-insecure people are often less likely to have physical access to markets where they can buy nutritious and healthy foods at affordable prices, particularly in high-income countries. The negative effect of food insecurity on diet quality has been documented in low-, middle- and high-income countries alike.[42]

There is a psychosocial route from food insecurity to obesity as well. The experience of not having certain or adequate access to food often causes feelings of anxiety, stress and depression, which in turn can lead to behaviours that increase the risk of overweight and obesity. These include patterns of binging or overeating when food is available (and continued availability uncertain), or choosing low-cost, energy-dense "comfort foods" rich in fat, sugar and salt. Such foods have been found to have physiological effects that reduce stress in the short term. As mentioned previously, the stress of living with food insecurity can also have a negative effect on breastfeeding and young child feeding practices, which in turn increases the children's risk of obesity in adulthood.[43]

Metabolic changes caused by disordered eating patterns and food deprivation are another component of the obesogenic pathway from food insecurity to malnutrition. Physiological adaptations in response to "feast-and-famine" cycles have been associated with an increase in body fat, decrease in lean muscle mass, and more rapid weight gain when food becomes plentiful.[44] In addition, maternal and infant/child food deprivation can result in foetal and early childhood "metabolic imprinting", which increases the risk of obesity and diet-related non-communicable chronic diseases later in life. Maternal undernutrition – as well as overweight – caused by lack of stable access to adequate diets can programme metabolic, physiological and neuroendocrine functions in offspring, fuelling an intergenerational cycle of malnutrition.[45]

As mentioned, food insecurity is associated with low birthweight in infants.[46] Low birthweight is a risk factor for child stunting, which in turn is associated with overweight and obesity later in life. According to a 2016 WHO report of the Commission on Ending Childhood Obesity: "Children who have suffered from undernutrition and were born with low birthweight or are short-for-age (stunted) are at far greater risk of developing overweight and obesity when faced with energy-dense diets and a sedentary lifestyle later in life."[47] It is also worth noting that children who are stunted have a higher risk of being simultaneously overweight.

Evidence on the association between poor access to food and obesity or overweight is growing, in resource-rich and resource-poor settings alike. In the context of the nutrition transition, overweight and obesity are not just problems for high-income countries, but increasingly an issue for low- and middle-income countries as well.

Evidence of the impact of food insecurity on malnutrition via the obesogenic pathway is especially notable in countries that have undergone the nutrition transition. Adult women who live in food-insecure households are at a higher risk of obesity, particularly in upper-middle- and high-income countries.[48] However, this link is weaker or absent for men, and there are no conclusive results for children, although food insecurity does appear to be associated with overweight in girls older than five.

According to the 2016 WHO report mentioned above, in high-income countries, childhood obesity is more prevalent among the lower socio-economic groups. The opposite is currently true in most low- and middle-income countries, although this pattern is rapidly changing. Indeed, certain subpopulation groups in these countries, such as indigenous populations, are at an even higher risk of becoming obese. In addition, according to the WHO report, "Childhood obesity is a strong predictor of adult obesity, which has well known health and economic consequences, both for the individual and society as a whole."[49]

In summary, there is little doubt that food insecurity is among the determinants of various forms of malnutrition via the pathways depicted in Figure 14. Food insecurity particularly increases the risk of low birthweight, stunting among

children under five and anaemia in women of reproductive age. It also interferes with exclusive breastfeeding of children in the first six months of life. Moreover, food insecurity is linked with overweight in girls older than five and is a risk factor for obesity among women, particularly in upper-middle- and high-income countries.

These findings, needless to say, are heavily dependent on context and research methods. Contextual factors such as country income level, or urban versus rural area, may explain some differences in the observed relationships between food insecurity and nutritional outcomes. Moreover, the majority of the studies are cross-sectional, meaning they did not involve observation of the same group over time. Longitudinal research, therefore, is necessary to understand the potential effects of food insecurity on nutritional outcomes throughout the life cycle, from before and during pregnancy to infancy and into adulthood.

It is equally important to analyse food insecurity at the individual level to highlight possible differences within households. The distribution of food and resources within households is influenced by a number of cultural and social factors. Especially under conditions of scarcity, women and children are sometimes discriminated against in the distribution of food; mothers may subsequently adjust their food intake to buffer the effect of food insecurity on their children. Gender inequalities in society and women's roles influence decision-making power and access to food within the household, with important consequences for women's own food security and nutrition as well as that of their children.

It is important to highlight that the experience of food insecurity also has other harmful consequences for the well-being of children and adults beyond malnutrition. Food insecurity has negative impacts on the academic performance of children and is associated with behavioural problems. Food-insecure children are more likely to face adverse health outcomes and developmental risks.[50] For children as well as adults, lack of reliable access to food can lead to anxiety, stress, depression, interpersonal tensions, and the alienation that comes with social stigma. These psychological and social effects have important consequences for overall health and well-being, regardless of nutritional status, and have negative economic impacts on individuals, households, communities and nations. They can contribute to a vicious cycle of food insecurity, wherein social isolation, depression and stress, as well as poor health and poor cognitive development, all impede people from reaching their full potential, with possible negative consequences for earning capacity and access to food.

So what can be done?

As more data become available on food security (food access), dietary intake and nutritional outcomes, integrated analysis of these data will yield better information to shape policies that address the multiple forms of malnutrition.

Existing evidence supports the need for implementing and scaling up interventions aimed at guaranteeing access to nutritious foods and breaking the intergenerational cycle of malnutrition. The 1 000 days between conception and a child's second birthday is a window of unsurpassed opportunity to both prevent child stunting and overweight and promote child nutrition, growth and development with lasting effects over the child's life. The origins of growth faltering begin as early as before and during pregnancy, with short- as well as long-term consequences. Child undernutrition can cause impaired cognitive development in children, with dramatic consequences in terms of self-realization and productivity. This can result in an intergenerational cycle of malnutrition, perpetuated by undernourished girls becoming undernourished mothers at risk of giving birth to infants with low birthweights. Exclusive breastfeeding in the first six months and adequate complementary foods and feeding practices up to two years of age are key to ensuring normal child growth and development during this crucial window of opportunity.

Given this evidence, policies must pay special attention to the food security and nutrition of infants and children under five, school-age children, adolescent girls and women. These groups have been identified as the most vulnerable to the harmful consequences of poor

food access. The ICN2 Framework for Action outlines relevant sets of recommended actions for improving food security and nutrition, which countries are committed to implementing under the umbrella of the UN Decade of Action on Nutrition.

"Double-duty actions" have been proposed by WHO that can simultaneously reduce undernutrition and overweight and obesity.[51] They highlight the need to be careful so that strategies to address undernutrition in early life do not exacerbate overweight and obesity later in life. Existing programmes should be redesigned and leveraged, and new interventions should be developed, to reduce the risk of multiple forms of malnutrition. Trade, investments and agriculture policies must be nutrition-sensitive and improve access to healthy diets, rather than promoting commodity crops that provide a cheap source of starch, fat and sugar in the food supply.[52]

The discussion illustrates why it is so important – especially in the context of the UN Decade of Action on Nutrition and the 2030 Agenda – to improve the way hunger and food insecurity are conceptualized and measured. Food insecurity can exist in all countries, and it can contribute to multiple forms of malnutrition – undernutrition and micronutrient deficiencies as well as overweight and obesity. Experience-based metrics of food insecurity like the FIES, and awareness of the different pathways from food insecurity to malnutrition, can contribute to the design of more effective interventions and policy coherence across sectors. The consequences for people's health, well-being and productivity are far-reaching.

In conclusion, evidence continues to point to a rise in world hunger and food insecurity in recent years. Progress is being made on child stunting – though too slow to meet global targets and with significant interregional and intraregional disparities. Simultaneously, rates of anaemia in women of reproductive age and obesity in adults are increasing. It will not be possible to end all forms of malnutrition without ensuring access to safe, nutritious and sufficient food all year round. This will require expanding the reach of social protection policies to address inequalities and ensuring that they are nutrition- and gender-sensitive in terms of: targeting; design; and in the identification of complementary health care and agriculture interventions to enhance nutrition outcomes. At the same time, a sustainable shift must be made towards nutrition-sensitive agriculture and food systems that can provide safe and high quality food for all promoting healthy diets in line with the recommended action of the ICN2 Framework for Action and the Work Programme of the UN Decade of Action on Nutrition.[53] Market regulations that discourage consumption of unhealthy foods are also called for, in conjunction with policies that promote the availability and consumption of healthy foods.[54] All of these actions require strengthened public governance and addressing conflicts of interest and imbalances in power among stakeholders. Access to food must be framed as a human right, prioritizing the access of the most vulnerable to safe, nutritious and sufficient food.

Part 2 takes an in-depth look at a factor that already appears to be having an impact on food security and nutrition, raising additional policy considerations: climate variability and extremes. ■

SAGAING REGION, MYANMAR
A rural woman benefitting from an FAO project to restore livelihoods and enhance resilience of disaster-affected communities in Myanmar.
©FAO/Hkun Lat

PART 2
THE IMPACT OF CLIMATE ON FOOD SECURITY AND NUTRITION

THE IMPACT OF CLIMATE ON FOOD SECURITY AND NUTRITION

As shown in Part 1 of this report, the number of people who suffer from hunger has been growing over the past three years, returning to levels that prevailed almost a decade ago. Equally of concern is that 22.2 percent of children under five are affected by stunting in 2017.

Last year this report observed that three factors are behind the recent trends affecting food security and nutrition in multiple ways and challenging people's access to food: conflict, climate and economic slowdowns. After an in-depth study of the role of conflict in the 2017 report, this part of the 2018 report focuses on the role of climate – more specifically, climate variability and extremes.

Here in Part 2, the report aims to understand how climate variability and extremes are adversely affecting food security and nutrition. The channels through which this is occurring are identified on the basis of existing evidence complemented with original analysis. The ultimate purpose is to provide guidance on how the key challenges brought about by climate variability and extremes can be overcome in order to achieve the goals of ending hunger and malnutrition in all forms by 2030 (SDG Targets 2.1 and 2.2) as well as other SDGs, including taking action to combat climate change and its impacts (SDG13).

2.1 WHY FOCUS ON THE IMPACT OF CLIMATE VARIABILITY AND EXTREMES ON FOOD SECURITY AND NUTRITION?

KEY MESSAGES

➔ Climate variability and exposure to more complex, frequent and intense climate extremes are threatening to erode and even reverse the gains made in ending hunger and malnutrition.

➔ Climate variability and extremes are a key driver behind the recent rise in global hunger and one of the leading causes of severe food crises.

➔ Severe droughts linked to the strong El Niño of 2015–2016 affected many countries, contributing to the recent uptick in undernourishment at the global level.

➔ Hunger is significantly worse in countries with agricultural systems that are highly sensitive to rainfall and temperature variability and severe drought, and where the livelihood of a high proportion of the population depends on agriculture.

Mounting evidence points to the fact that climate change is already affecting agriculture and food security, which will make the challenge of ending hunger, achieving food security, improving nutrition and promoting sustainable agriculture more difficult.[55]

FIGURE 15
INCREASING NUMBER OF EXTREME CLIMATE-RELATED DISASTERS, 1990–2016

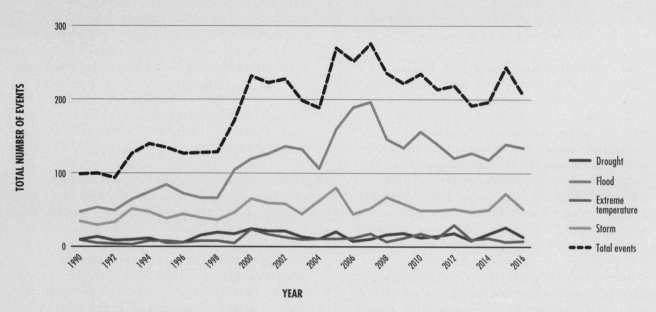

NOTE: Total number of natural disasters that occurred in low- and middle-income countries by region and during the period 1990–2016. Disasters are defined as medium- and large-scale disasters that exceed the thresholds set for registration on the EM-DAT international disaster database. See Annex 2 for the full definition of EM-DAT disasters.
SOURCE: FAO elaboration based on data from Emergency Events Database (*EM-DAT*). 2009. *EM-DAT* [online] Brussels. www.emdat.be

Climate change takes place over a period of decades or centuries. There are also shorter-term climate variations (e.g. in temperature and rainfall) and extremes (leading to drought, floods, storms, etc.) associated with periodic or intermittent changes related to different natural phenomena (such as El Niño, La Niña, volcanic eruptions or other changes in earth systems).[56] However, these shorter-term climate variations are not all attributable to climate change.

In any case, the *attribution* of climate variations and extremes to climate change is beyond the scope of this report.

The focus on climate variations and extremes is prompted by three considerations. *First*, the number of extreme events, including extreme heat, droughts, floods and storms, has doubled since the early 1990s, with an average of 213 of these events occurring every year during the period of 1990–2016 (Figure 15). *Second*, while

climate change occurs over a period of decades or centuries, what people experience in their daily life is climate variability and climate extremes,[57] regardless of whether or not these are driven by climate change. *Third*, unsurprisingly, all dimensions of food security and nutrition, including food availability, access, utilization and stability, are potentially affected even in the short term by climate variability and climate extremes.

Changes in climate are already undermining production of major crops (wheat, rice and maize) in tropical and temperate regions and, without adaptation, this is expected to worsen as temperatures increase and become more extreme.[58] Climate-related disasters have come to dominate the risk landscape to the point where they now account for more than 80 percent of all major internationally reported disasters.[59] Of all natural hazards, floods, droughts and tropical storms affect food production the most. Drought in particular causes more than 80 percent of the total damage and losses in agriculture, especially for the livestock and crop production subsectors. In relation to extreme events, the fisheries subsector is most affected by tsunamis and storms, while most of the economic impact on forestry is caused by floods and storms.[60]

New information from country food balance sheets points to reductions in food availability and price increases in regions affected by the El Niño phenomenon in 2015–16. This event resulted in large climatic deviations and anomalies compared to historical norms, which were experienced in different ways and to varying degrees of intensity in various parts of the world (Box 8). In some areas, severe drought conditions have resulted from the El Niño phenomenon, particularly in regions where many low- and middle-income countries are situated.

While hunger is on the rise, it is equally alarming that the number of people facing crisis-level food insecurity continues to increase. In 2017, almost 124 million people across 51 countries and territories faced "crisis"[61] levels of acute food insecurity or worse, requiring immediate emergency action to safeguard their lives and preserve their livelihoods.[62] This represents an increase compared to 2015 and 2016, when 80 and 108 million people, respectively, were reported as facing crisis levels. As with increased levels of hunger, major contributors to crisis-level food insecurity are climate-related, in particular droughts. Moreover, climate variability and extremes are also contributing to the alarming levels of malnutrition, as can be seen below.

The 2030 Agenda: advancing progress through strengthened resilience and adaptive capacity in response to natural hazards and climate-related disasters

The 2030 Agenda for Sustainable Development makes an explicit link between sustainable development and climate action. Through SDG13, the 2030 Agenda calls for strengthened resilience and adaptive capacity in response to natural hazards and climate-related disasters in all countries.[63] It also calls on all countries to establish and operationalize an integrated strategy – one that includes food security and nutrition – to improve their ability to adapt to the adverse

impacts of climate change, and to foster climate resilience and lower greenhouse gas (GHG) emissions without jeopardizing food production.

Agricultural production and food systems are major sources of GHG emissions and are particularly sensitive to climate. These systems need to be a priority for climate change adaptation and mitigation action. The challenge is to increase agricultural production in ways that are both more sustainable (for example, through enabling sustainable healthy diets) and more climate-resilient, while at the same time reducing emissions.

Addressing climate variability and extremes and their impact on food security and nutrition requires cross-sectoral action with stakeholder engagement at all levels. A challenge is that existing global policy strategies are compartmentalized into several dialogues: climate change, governed by the United Nations Framework Convention on Climate Change (UNFCCC) and the 2015 Paris Agreement; disaster risk reduction, under the Sendai Framework for Disaster Risk Reduction; and the humanitarian–development nexus and resilience building, broadly addressed in the 2016 World Humanitarian Summit and subsequent discussions.

At the same time, nutrition, health and the links between them – which are all impacted by climate variability and extremes – are addressed in the outcome documents of the second International Conference on Nutrition (ICN2), where countries recognized the need to act. The Work Programme of the UN Decade of Action on Nutrition provides a framework for helping countries to implement relevant commitments and recommendations.

Similarly, these global policy dialogues are elaborated in a number of national action plans related to climate change, disaster risk reduction and resilience and nutrition. These include National Adaptation Plans (NAPs), Health National Adaptation Plans (HNAPs) and Nationally Determined Contributions (NDCs), which guide national climate change adaptation and mitigation action. The HNAPs usually include food and nutrition security. **»**

BOX 8
THE RELATIONSHIP BETWEEN CLIMATE VARIABILITY AND ENSO

The El Niño–Southern Oscillation (ENSO), the North Atlantic Oscillation (NAO) and the Indian Ocean Dipole (IOD) are among the large-scale drivers that combine to influence regional atmospheric circulation patterns, regional-scale drivers such as sea surface temperatures (SST), local drivers such as soil moisture conditions, and local stochastic effects such as the random location and track of a thunderstorm/cyclone over a region.

ENSO is one of Earth's most important climatic phenomena. The ENSO cycle describes the fluctuations in temperature between the ocean and atmosphere in the east-central Equatorial Pacific Ocean. La Niña is known as the cold phase and El Niño as the warm phase of ENSO. These temperature variations can have large-scale impacts not only on ocean processes, but also on global weather and climate. As shown in the figures below, El Niño commonly results in impacts on different regions of the globe and during different seasons.

The 2015–2016 El Niño was extreme and one of the strongest events of the past 100 years. It resulted in

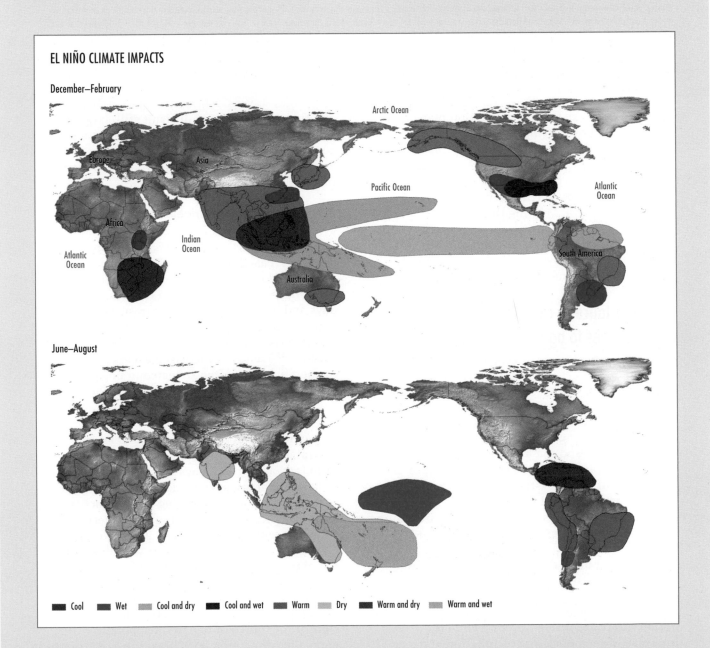

EL NIÑO CLIMATE IMPACTS

December–February

June–August

Cool | Wet | Cool and dry | Cool and wet | Warm | Dry | Warm and dry | Warm and wet

NOTES: The final boundary between the Republic of the Sudan and the Republic of South Sudan has not yet been determined. Final status of the Abyei area has not yet been determined.

SOURCE: Weather Impacts of ENSO (available at www.weather.gov/jetstream/enso_impacts).

**BOX 8
(CONTINUED)**

record-breaking warm conditions for many tropical and subtropical countries: 2015 and 2016 were two of the warmest years on record for global average surface air temperature. Large parts of Asia and the Pacific experienced hot spring and summer seasons, and many extreme climate events were observed, including cyclones, flooding, severe droughts and extreme temperatures.

Sources: NOAA Climate.gov; C. Holleman, F. Rembold and O. Crespo (forthcoming). *The impact of climate variability and extremes on agriculture and food security: an analysis of the evidence and case studies.* FAO Agricultural Development Economics Technical Study 4. Rome, FAO; S. Hu and A.V. Fedorov. 2017. The extreme El Niño of 2015–2016 and the end of global warming hiatus. *Geophysical Research Letters*, 44(8): 3816–3824; B. Huang, M. L'Heureux, Z.-Z. Hu and H.-M. Zhang. 2016. Ranking the strongest ENSO events while incorporating SST uncertainty. *Geophysical Research Letters*, 43(17): 9165–9172.

» All these policy dialogues and action plans aim to achieve the overarching goal of sustainable development that is embodied in the 2030 Agenda. The challenge is to use policy and cross-sectoral strategies to strengthen resilience and adaptive capacities to climate variability and extremes (SDG13). Meeting this challenge through integrated solutions is absolutely necessary to end extreme poverty and hunger, achieve food security, improve nutrition, and make agriculture sustainable (SDGs 1 and 2).

The importance of changing climate variability and extremes to agriculture, food security and nutrition

There is strong evidence of global climate change in the form of rising air and sea surface temperatures, receding glaciers, shifting climate regimes, increasing frequency and intensity of extreme events and sea level rises.[64] The accelerated warming of the planet continues to lead to modified ecosystem processes, changing climate variability and more intense climate-related events across the globe, including extreme temperatures (cold and hot spells) and variations in rainfall (floods and droughts). However, as noted, not all types of climate and temperature extremes are easily attributable to climate change. For example, droughts are sometimes difficult to connect to warming trends because they are influenced by a complex interplay of temperature, precipitation and soil moisture, with precipitation in particular exhibiting high natural variability. Hurricanes

and typhoons are more difficult still, largely as they occur so rarely, and their dynamics are so complex. What is clear is that people are experiencing climate variability and extremes in their daily lives.

Increasing and more variable temperatures

The Earth's climate has experienced rapid warming of approximately 0.85 °C during the last century.[65] Based on historical observations, there is a clear global trend of an overall increase in warm days and nights, with a reduction in cold days and nights. Land and ocean surface temperature have clearly been rising over time and this rise has been accelerating in the last few decades.[66] Trends in increased average temperatures are often reflected in one or more measures of extreme temperature (e.g. hot/cold days and hot/cold nights).

In Australia, southern Africa, and northern, central, eastern and western Asia, there have been increases in hot days and hot nights. Even so, a few subregions have demonstrated spatially variable warming and cooling trends, such as eastern Africa, western and south-eastern South America, central North America and the eastern United States of America, along with decreases in hot nights in north-eastern Canada. Overall, in the northern hemisphere, 1983–2012 was the warmest 30-year period of the last 1 400 years.[67] Most recently, the 2015–2016 El Niño was a significant source of regional temperature anomalies, including both higher (e.g. Brazil) and lower (e.g. Kenya, the United Republic of Tanzania) surface temperatures.[68]

FIGURE 16
RECENT PAST TEMPERATURE ANOMALIES COMPARED TO THE 1981–2016 AVERAGE

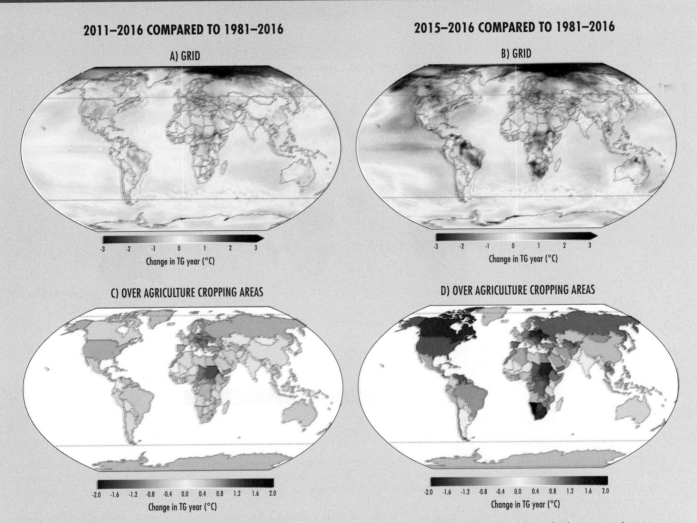

2011–2016 COMPARED TO 1981–2016

A) GRID

2015–2016 COMPARED TO 1981–2016

B) GRID

Change in TG year (°C)

C) OVER AGRICULTURE CROPPING AREAS

D) OVER AGRICULTURE CROPPING AREAS

Change in TG year (°C)

NOTES: The maps show changes in mean surface air temperature (TG) in degrees Celsius (°C). Figures 16a and 16b are grid-level figures. Figures 16c and 16d are aggregated per country over agriculture cropping areas. In these cases, climate data are given larger weight where there is cropping compared to where there is not. Areas with insufficient data coverage are denoted in grey. The final boundary between the Republic of the Sudan and the Republic of South Sudan has not yet been determined. Final status of the Abyei area has not yet been determined.
SOURCE: C. Holleman, F. Rembold and O. Crespo (forthcoming). *The impact of climate variability and extremes on agriculture and food security: an analysis of the evidence and case studies.* FAO Agricultural Development Economics Technical Study 4. Rome, FAO.

The temperature anomalies associated with El Niño serve to show that climate variability and extremes affect agriculture. The period 2015–2016 witnessed the most recent El Niño event, and 2011–2016 witnessed the longest recent time span with only one El Niño event (the previous was in 2010); both also align with the noted increase in PoU in many areas of the world. Hence it is useful to describe climate anomalies over these two periods in order to

unpack the possible links between climate and increasing PoU.

It can be noted that average temperatures over agriculture cropping areas are higher in most countries during both periods, compared with the long-term average of 1981–2016 (Figure 16). Where this occurs, there has likely been an impact on crop yields and production. There are some exceptions however: Argentina, Kenya, Paraguay,

FIGURE 17
NUMBER OF YEARS WITH FREQUENT HOT DAYS OVER AGRICULTURE CROPPING AREAS (2011–2016 COMPARED TO 1981–2016)

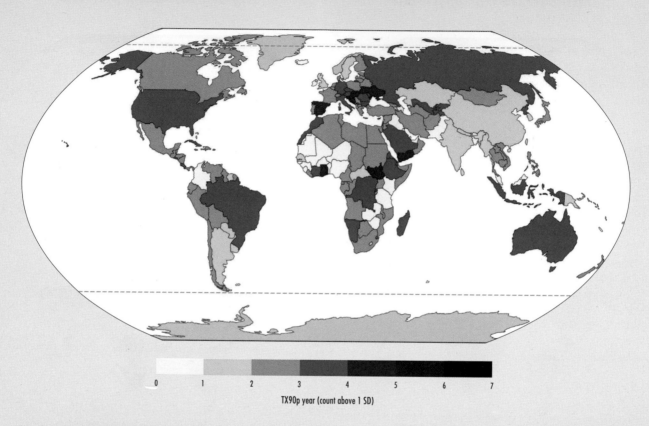

TX90p year (count above 1 SD)

NOTES: The map shows the number of years where the percentage of days when daily temperature is higher than the 90th percentile (TX90p) exceeds one annual standard deviation (SD). It uses country aggregate maximum temperature data over agriculture cropping areas. In these cases, climate data are given larger weight where there is cropping compared to where there is not. Areas with insufficient data coverage are denoted in grey. The final boundary between the Republic of the Sudan and the Republic of South Sudan has not yet been determined. Final status of the Abyei area has not yet been determined.
SOURCE: C. Holleman, F. Rembold and O. Crespo (forthcoming). *The impact of climate variability and extremes on agriculture and food security: an analysis of the evidence and case studies.* FAO Agricultural Development Economics Technical Study 4. Rome, FAO.

the United Republic of Tanzania, and parts of West Africa during 2015–2016, along with India, Pakistan, Indonesia and Malaysia during 2011–2016, all experienced cooler temperatures that may be related in some cases to increases in rainfall associated with El Niño.

In many areas, extremes have increased in number and intensity, particularly where average temperatures are shifting upwards: very hot days are becoming more frequent and the hottest days are becoming hotter. Extreme heat is associated with increased mortality, lower labour capacity, lower crop yields and other consequences that undermine food security and nutrition.

Temperature anomalies over agriculture cropped areas continued to be higher than the long-term mean throughout 2011–2016, leading to more frequent extremely hot conditions in the last five years (Figure 17). Many countries, including Brazil, Ethiopia, Indonesia and some others in East Africa and Central Asia have experienced three or more years where maximum daily temperatures were much more frequently extreme.

High spatial variability in rainfall

Annual precipitation (or rainfall) is naturally far more variable from year to year than temperature, and a range of drivers from local to global are responsible for this. Total rainfall changes depend on the variations in both frequency and intensity, which can either compensate or reinforce each other. For example, between 2011 and 2016, increases in rainfall frequency over Central Asia were compensated by decreases in intensity, whereas over Southern Africa both frequency and intensity declined.[69] Also, historical trends in rainfall are far more diverse depending on the region, although there seem to have been more regional increases than decreases in heavy rainfall.

Recent years show large spatial variability in rainfall data, displaying both strong positive and negative anomalies when compared with the historic average (Figure 18). Most notable are the below-normal rainfall levels over a large area of the globe in 2015–2016, some of which are also evident during the 2011–2016 period – again highlighting the influence of climate variability (especially strong global events such as ENSO) on the sub-decadal periods in which they occur. These anomalies are also apparent when aggregated over agriculture cropping areas (Figure 18c, d), which is equally striking, with below-normal precipitation levels during 2015–2016 in Africa, Central and South America, South-eastern Asia, the Philippines and Papua New Guinea. These are regions where the livelihoods of millions of small-scale family farmers, pastoralists and agropastoralists depend on rainfall – but above-normal rainfall is often hazardous and leads to crop damage, soil erosion and flooding. During the 2015–2016 El-Niño, large parts of Asia experienced higher than normal rainfall.

Changes in seasonality

In addition to increasing temperatures and changes in rainfall, the nature of rainfall seasons is also changing, specifically the timing of seasonal climate events. This is related to the late/early start of rainy seasons, the unequal distribution of rainfall within a season (e.g. periods of dry and rainy days) and changes in temperatures during the rainfall season. Within-season changes may not register as extreme climate events (droughts, floods or storms) but rather are aspects of climate variability that affect the growth of crops and the availability of pasture for livestock, with potentially significant implications for food security and nutrition.

For example, in the Afram Plains region of Ghana, farmers are noticing delays in the onset of the rainy season, mid-season heatwaves and high-intensity rains that cause flooding, which are resulting in crop loss, low yields and reduced availability of household food.[70] Similarly in Wenchi, Ghana, farmers consider both poor rainfall distribution and frequent droughts as the most important climate-related changes.[71] Farmers in the Nigerian savannah and the Kagera region in the north of the United Republic of Tanzania are also noticing changing rainfall patterns and shorter growing seasons.[72] Very few studies, however, have linked farmer reports of changing seasonal patterns to actual climatic data.[73]

It is difficult to understand the causes and impacts of changes in seasonal rainfall distributions, lengths of seasons and start/end of seasons, as these depend on the specific crop and livestock system, as well as the multitude of differing agriculture calendars. However, the frequency and intensity of daily rainfall (see Figure 20) provide some evidence that many countries and regions have experienced changes in the distribution of rainfall over cropped areas in the last few years.

Africa is one region where the influence of climate on production and livelihoods is both strongest and most complex. Much of the vulnerability to climate shocks stems from the dryland farming and pastoral rangeland systems that dominate livelihood systems for 70–80 percent of the continent's rural population.[74] A heavy reliance on rainfed agriculture (crops and rangelands) makes rural populations more vulnerable. Furthermore, in arid, semi-arid and dry subhumid areas, the impacts of human activities aggravate conditions of desertification and drought. This is particularly relevant to Africa as farming

FIGURE 18
RECENT PAST PRECIPITATION ANOMALIES COMPARED TO THE 1981–2016 AVERAGE

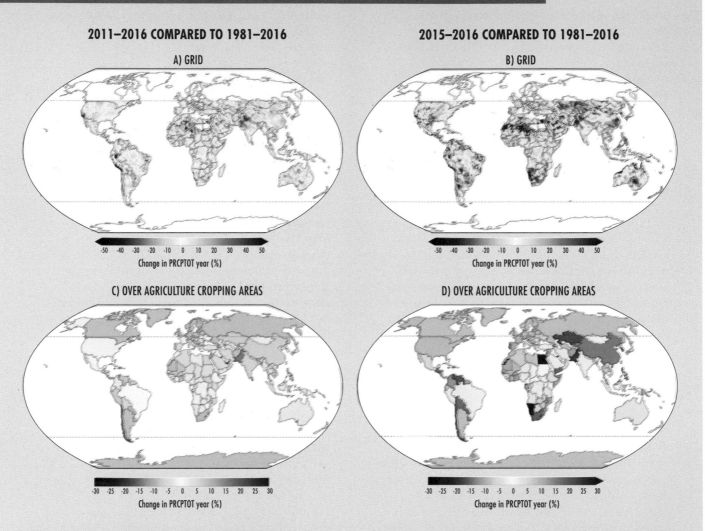

2011–2016 COMPARED TO 1981–2016

A) GRID

2015–2016 COMPARED TO 1981–2016

B) GRID

Change in PRCPTOT year (%)

C) OVER AGRICULTURE CROPPING AREAS

D) OVER AGRICULTURE CROPPING AREAS

Change in PRCPTOT year (%)

NOTES: Comparison of average annual precipitation (PRCPTOT) anomalies. The relative changes in precipitation in Figures 18c,d are aggregated per country over agriculture cropping areas. In these cases, climate data are given larger weight where there is cropping compared to where there is not. Areas with insufficient data coverage are denoted in grey. The final boundary between the Republic of the Sudan and the Republic of South Sudan has not yet been determined. Final status of the Abyei area has not yet been determined.
SOURCE: C. Holleman, F. Rembold and O. Crespo (forthcoming). *The impact of climate variability and extremes on agriculture and food security: an analysis of the evidence and case studies.* FAO Agricultural Development Economics Technical Study 4. Rome, FAO.

practices are extended into agriculture on marginal lands (e.g. arid and semi-arid lands, hilly and mountainous areas and wetlands).[75] The strength and complexity of the relationships to climatic influences in this region, coupled with one of the highest prevalence rates of undernourishment and undernutrition in the world, warrants a more in-depth analysis to detect changes in the length and onset of seasons.

Figure 19 shows the major emerging trends in cropland and rangeland vegetation growing season length (GSL) in Africa between 2004 and 2017. The left panel indicates that GSL was significantly reduced in western and southern Africa (red colours). The colour scale of the right panel indicates which year was most extreme in terms of (smaller) vegetation production. Altogether, the figure reveals some spatial patterns. For example, in many

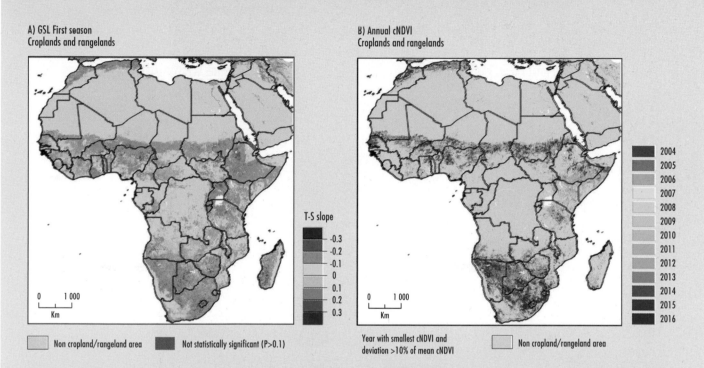

FIGURE 19
DECREASED GROWING SEASON LENGTH AND YEAR OF LOWEST CUMULATIVE ANNUAL VEGETATION BIOMASS OVER CROPLAND AND RANGELAND AREAS IN AFRICA, 2004–2016

A) GSL First season
Croplands and rangelands

B) Annual cNDVI
Croplands and rangelands

T-S slope

-0.3
-0.2
-0.1
0
0.1
0.2
0.3

2004
2005
2006
2007
2008
2009
2010
2011
2012
2013
2014
2015
2016

Non cropland/rangeland area   Not statistically significant (P>0.1)

Year with smallest cNDVI and deviation >10% of mean cNDVI   Non cropland/rangeland area

NOTES: Figure 19a shows cropland and rangeland vegetation growing season length (GSL) trends. The orange to red colours identify areas with significantly reduced length of growing season. Figure 19b shows the year with the lowest annual vegetation biomass production based on remote sensing vegetation coverage data, represented through the annual cumulative value of the Normalized Difference Vegetation Index (cNDVI). The colour scale indicates which year was most extreme in terms of minimum vegetation production. T-S slope is the average change in dekad (10-day period) per year. The final boundary between the Republic of the Sudan and the Republic of South Sudan has not yet been determined. Final status of the Abyei area has not yet been determined.
SOURCE: C. Holleman, F. Rembold and O. Crespo (forthcoming). *The impact of climate variability and extremes on agriculture and food security: an analysis of the evidence and case studies.* FAO Agricultural Development Economics Technical Study 4. Rome, FAO.

countries in southern Africa (Angola, Botswana, Lesotho, Madagascar, Malawi, Namibia and South Africa), blue areas suggest that the El Niño period 2015–2016 had the poorest production. The same applies to parts of northern Africa, which experienced a major drought in 2016. Furthermore, 2011 was the year with the poorest growing season for a significant part of eastern Africa, which experienced a major drought in that period following the 2010 La Niña. The period 2004–2005 also witnessed many droughts across the continent, with minimal biomass production in many regions.

Severe droughts
Droughts are extreme climate events characterized by prolonged periods of rainfall deficits that can result in food insecurity and malnutrition, largely through cascading negative effects on agriculture production, food prices, value chains, water supplies and livelihoods, affecting access to income and food.

Evidence shows that recent years (2011–2016) have been characterized by a number of severe droughts in many regions. Some of these feature among the most extreme droughts historically (e.g. state of California in the United States of »

FIGURE 20

PRECIPITATION ANOMALIES ASSOCIATED WITH DROUGHT IN AGRICULTURE CROPPING AREAS (2011–2016 COMPARED TO 1981–2016)

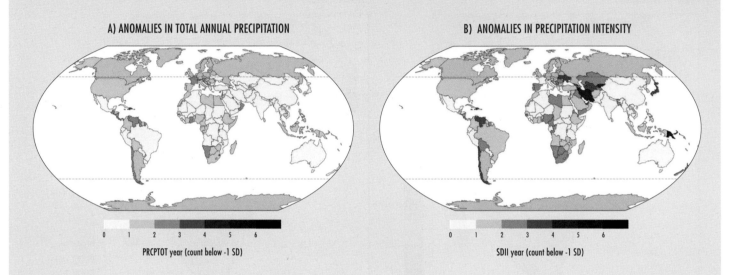

A) ANOMALIES IN TOTAL ANNUAL PRECIPITATION

PRCPTOT year (count below -1 SD)

B) ANOMALIES IN PRECIPITATION INTENSITY

SDII year (count below -1 SD)

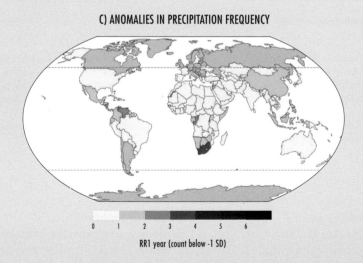

C) ANOMALIES IN PRECIPITATION FREQUENCY

RR1 year (count below -1 SD)

NOTES: The maps show the number of years a country experienced negative precipitation anomalies in the period 2011–2016 in terms of: total accumulated rainfall in year as measured by total annual precipitation (PRCPTOT) (Figure 20a); rainfall intensity as measured by the ratio of annual total rainfall to the number of days during the year when rainfall occurred (SDII) (Figure 20b); and precipitation frequency as measured by the number of days when rainfall was above 1 mm (RR1) (Figure 20c). More than three years of occurrence out of seven for the period 2011–2016 is considered outside normal variation (below - 1 standard deviation [SD]). Country climate data are aggregated over cropping areas smoothed for small geographical scale events, especially in large countries. Areas with insufficient data coverage are denoted in grey. The final boundary between the Republic of the Sudan and the Republic of South Sudan has not yet been determined. Final status of the Abyei area has not yet been determined.

SOURCE: C. Holleman, F. Rembold and O. Crespo (forthcoming). *The impact of climate variability and extremes on agriculture and food security: an analysis of the evidence and case studies.* FAO Agricultural Development Economics Technical Study 4. Rome, FAO.

» America; Australia), while others were unusually prolonged and spread over larger areas (e.g. Somalia, Southern Africa, India and the Dry Corridor of Central America).[76]

Counting the number of years when large precipitation deficits were observed during the last five years (Figure 20a) shows that a number of countries experienced large negative rainfall anomalies with higher frequency during the period 2011–2016, compared with the longer period of 1981–2016. Several countries – notably in Africa, Central America and South-eastern Asia – experienced drought, not only through abnormally low total accumulated rainfall (Figure 20a), but also through lower rainfall intensities and fewer days of rainfall (Figure 20b, c).

Significantly lower frequency and smaller amounts of precipitation for larger areas lead to drought, which is particularly worrying for agricultural production. The duration of a drought is often a critical factor in its overall impact on food security and nutrition. Indicators of frequency and duration include, among others, rainfall deficits and normalized difference vegetation index (NDVI) anomalies during growing seasons.[77] Globally, the years 2004–2006 and 2015 registered the highest frequency of drought conditions for crops since the mid-2000s, coinciding with ENSO anomalies (El Niño in 2004–2005, 2006–2007 and 2015–2016). The same data suggest that 2009 and 2011 were also important drought years, for example in large parts of eastern Africa.[78]

The impact of the 2015–2016 El Niño on agricultural vegetation is clearly visible when comparing the frequency of drought conditions in 2015–2017 with those of 2004–2017 (Figure 21). The 2015–2017 map shows that large areas in Africa, parts of Central America, Brazil and the Caribbean, as well as Australia and parts of the Near East experienced a large increase in frequency of drought conditions in 2015–2017 compared to the 14-year average. Although there is regional variability, since the end of the 1960s the Sahel, the Horn of Africa, and southern Africa have been particularly affected by

drought.[79] These have led to severe famine and socio-economic losses (e.g. loss of livestock) as well as an increase in disease and illness.

Severe floods and storms

Floods cause more climate-related disasters globally than any other extreme climate event, with flood-related disasters seeing the highest increase – 65 percent – in occurrence over the last 25 years (Figure 22a). Asia is the region with the highest occurrence of flood disasters. However, flood-related disasters in Africa have declined dramatically since 2006 and were surpassed by those in Latin America and the Caribbean in 2013.

The frequency of storms is not increasing as much as that of floods (Figure 15), but storms are the second most frequent driver of climate-related disasters. Storm-related disasters are again highest in Asia, averaging between 20 to 30 every year (Figure 22b). Some parts of Africa also register a high number of storm-related disasters, but these tend to be more localized.

River floods, oceanic storm surges and tropical cyclones negatively impact low-lying areas, flood plains and deltas. A detailed study of 33 deltas around the world found that 85 percent had experienced severe flooding in the past decade, affecting an area of 260 000 km². [80]

Although flood- and storm-related disasters have generally increased in number over time, fewer people are now affected by them. An analysis of annual fatalities from tropical cyclones revealed these to be heavily concentrated in low-income nations, though there was high exposure in many upper-middle- and high-income nations as well (with larger economic losses in these nations).[81]

A regional analysis of changes in exposure, vulnerability and risk indicates that although exposure to floods and cyclones has increased since 1980, the risk of mortality has generally fallen.[82] Nonetheless, evidence suggests that food insecurity and malnutrition risks are magnified due to the high vulnerability of agriculture, food systems and livelihoods to climate extremes including floods and storms (see next section).

FIGURE 21
FREQUENCY OF AGRICULTURAL DROUGHT CONDITIONS DURING THE EL NIÑO OF 2015–2017 COMPARED TO THE 2004–2017 AVERAGE

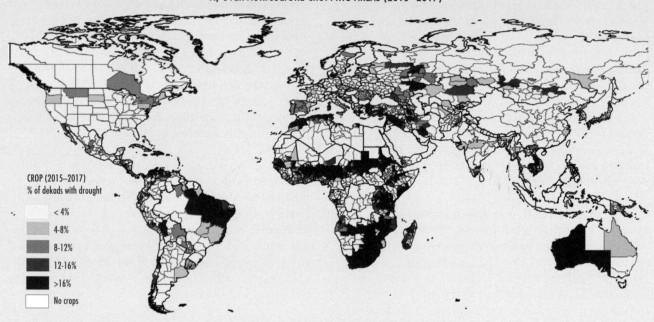

A) OVER AGRICULTURE CROPPING AREAS (2015–2017)

CROP (2015–2017)
% of dekads with drought

- < 4%
- 4-8%
- 8-12%
- 12-16%
- >16%
- No crops

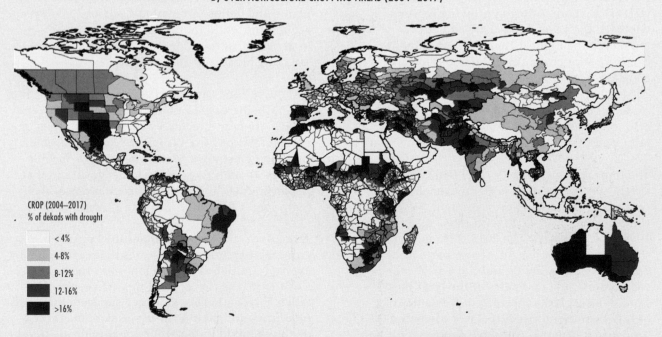

B) OVER AGRICULTURE CROPPING AREAS (2004–2017)

CROP (2004–2017)
% of dekads with drought

- < 4%
- 4-8%
- 8-12%
- 12-16%
- >16%

NOTES: Figure 21 shows the percentage of time (dekad is a 10-day period) with active vegetation when the Anomaly Hot Spots of Agricultural Production (ASAP) was signalling possible agricultural production anomalies according to NDVI (drought warning) for more than 25 percent of the crop areas in 2015–2017 (a) compared to 2004–2017 (b). The final boundary between the Republic of the Sudan and the Republic of South Sudan has not yet been determined. Final status of the Abyei area has not yet been determined.
SOURCE: ASAP early warning system; European Commission Joint Research Centre (EC-JRC); C. Holleman, F. Rembold and O. Crespo (forthcoming). *The impact of climate variability and extremes on agriculture and food security: an analysis of the evidence and case studies.* FAO Agricultural Development Economics Technical Study 4. Rome, FAO.

FIGURE 22
FREQUENCY OF FLOOD- AND STORM-RELATED DISASTERS BY REGION, 1990–2016

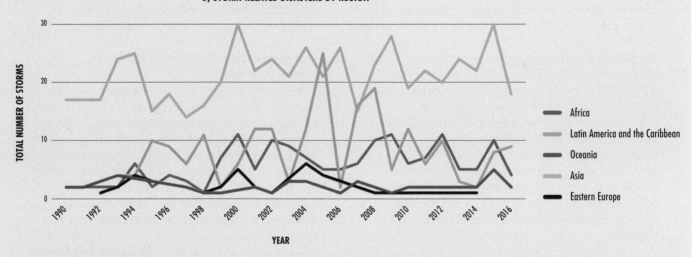

A) FLOOD-RELATED DISASTERS BY REGION

B) STORM-RELATED DISASTERS BY REGION

NOTES: Total number of flood- (Figure 22a) and storm-related (Figure 22b) disasters that occurred in low- and middle-income countries by region and during the period 1990–2016. Disasters are defined as medium- and large-scale disasters that exceed the thresholds set for registration on the EM-DAT international disaster database. See Annex 2 for the full definition of EM-DAT disasters.
SOURCE: FAO elaboration based on data from Emergency Events Database (EM-DAT). 2009. *EM-DAT* [online] Brussels. www.emdat.be

Climate impacts on food security and nutrition

Both climate variability and extremes have implications for agriculture and food production. As a result, all dimensions of food security and nutrition are likely to be affected, including food availability, access, utilization and stability. The association between climate variability and extremes and food security and nutrition indicators corroborates this.

Increases in undernourishment associated with severe drought

Food security and nutrition indicators can be associated particularly with an extreme climate event, such as a severe drought, that critically challenges agriculture and food production. If a drought is severe and widespread enough, it can potentially affect national food availability and access, as well as nutrition, thus magnifying the prevalence of undernourishment (PoU) nationally.[83] This is particularly the case where a country's agricultural production is highly vulnerable to climate variability and extremes and the country does not have in place sufficient support measures to counter the fallout.

Although it is difficult to establish a direct causal relationship considering the way the PoU is computed and smoothed over time,[84] it is possible to examine whether change points in the PoU time series correspond to occurrences of severe drought.[85]

A change point analysis of PoU time series, identifying years of increasing undernourishment after years of reduction or stabilization, indicates that out of 91 PoU change points in 76 countries, 28 of them in 27 countries occurred in correspondence with severe drought stress conditions between 2006 and 2016 (see Annex 3 for methodology). In other words, for almost 36 percent of the countries that experienced a rise in undernourishment since 2005, this coincided with the occurrence of severe drought. Out of 27 countries with change points occurring under severe drought stress conditions, most (19 countries) are in Africa, with the remaining 4 in Asia, 3 in Latin America and the Caribbean, and 1 in Eastern Europe (Figure 23).

Most striking is the significant increase in the number of change points related to severe drought conditions in 2014–2015, in which nearly two-thirds of the change points occurred. In these cases, the PoU increased from 2015 onwards and this can be linked to severe droughts driven by El Niño in 2015–2016. A closer review reveals that many countries have witnessed periods of increased undernourishment over the past years; however, during the period of the ENSO event of 2015–2016, this change across so many countries contributed to a reversal of the PoU trend at the global level.

Although the analysis is not causal and data limitations prevent a statistical inference of association, the outcome of so many corresponding occurrences suggests that drought could be one important contributing factor to the recent PoU increases in some cases. This change point analysis does support the hypothesis that – particularly for the period of 2014–2016 – extreme drought linked to the strong El Niño of 2015–2016 is one of the drivers behind the increases in PoU. This association is further corroborated by a number of studies that show a strong link between drought and stunting in children. For example, drought events in Bangladesh are associated with a higher stunting rate around five and nine months after the beginning of the drought event.[86] In rural Zimbabwe, one- to two-year olds exposed to drought face significantly lower growth velocity compared to children of the same age living in areas with average rainfall.[87] In sub-Saharan Africa, warmer and drier climates are related to declining food availability and increased prevalence estimates of childhood stunting.[88]

Increased exposure and vulnerability to climate extremes

The extent to which climate variability and extremes negatively affect people's food security and nutrition situation depends on their degree of exposure to climate shocks and vulnerability to these shocks. In the analysis that follows, climate shocks are defined as the occurrence of extreme rainfall and/or temperatures over agriculture areas but also complex events (e.g. droughts, storms and floods) in each year of a given timeframe. In the last 20 years, not only has exposure to climate shocks risen in terms of both frequency and intensity, but this has occurred in countries already vulnerable to the risk of food insecurity and malnutrition. Specifically, there has been an increase in climate shocks caused by drought, floods, storms and heat spells in countries where undernourishment, production and yields are vulnerable to climate extremes.

Looking at country exposure to climate extremes, evidence indicates that the number of low- and middle-income countries exposed to climate extremes has increased, from 83 percent of countries in 1996–2000 to 96 percent in

FIGURE 23
PoU CHANGE POINTS ASSOCIATED WITH THE OCCURRENCE OF SEVERE AGRICULTURAL DROUGHT

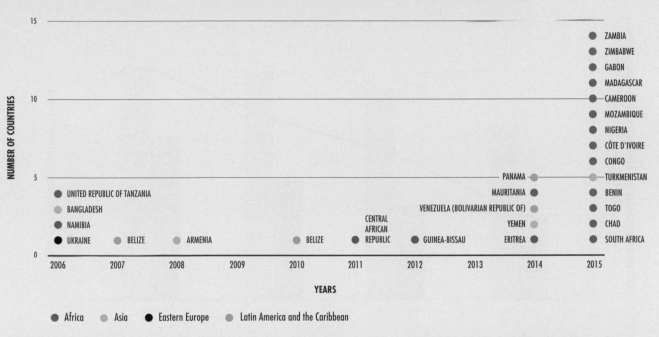

NOTE: The number of countries with change points of prevalence of undernourishment (PoU) which occurred in correspondence with severe drought conditions by year, between 2006 and 2015. See Annex 3 for methodology and list of countries with PoU change points related to severe drought conditions.
SOURCE: C. Holleman, F. Rembold and O. Crespo (forthcoming). *The impact of climate variability and extremes on agriculture and food security: an analysis of the evidence and case studies.* FAO Agricultural Development Economics Technical Study 4. Rome, FAO.

2011–2016 (Figure 24). Most striking is that the frequency (number of years exposed in a five-year period) and intensity (multiple types of climate extremes in a five-year period) of exposure to climate extremes have both increased too. Considering the frequency, or number of years exposed in each subperiod, countries' exposure increased by more than 30 percent between 1996–2000 and 2011–2016. In terms of increasing intensity, 36 percent of countries were exposed to three or four types of climate extremes (extreme heat, drought, floods or storms) in 2011–2016, compared with 18 percent in 1996–2000. In other words, the number has doubled in the last 20 years (see Annex 2 for definitions and methodology).

Looking at the regional level, the analysis reveals even greater increases in the intensity of climate

extremes compared to the global averages. For instance, the occurrence of three or more different types of climate extremes has increased by 160 percent for countries in Africa, from 10 percent in 1996–2000 to 25 percent in 2011–2016. Similarly, the percentage of Asian countries experiencing multiple shocks more than doubled to 51 percent in 2011–2016, up from 23 percent in 1996–2000. The intensity of climate extremes in Latin America and the Caribbean also more than doubled (from 26 percent in 1996–2000 to 56.5 percent in 2011–2016).

Many countries – especially in Africa and Asia – are also now more exposed to interseasonal climate variability, either in terms of early or delayed onset of growing seasons, decreased length of the growing seasons, or both. Fifty-one low- and middle-income countries experienced

FIGURE 24
INCREASED EXPOSURE TO MORE FREQUENT AND MULTIPLE TYPES OF CLIMATE EXTREMES IN LOW- AND MIDDLE-INCOME COUNTRIES

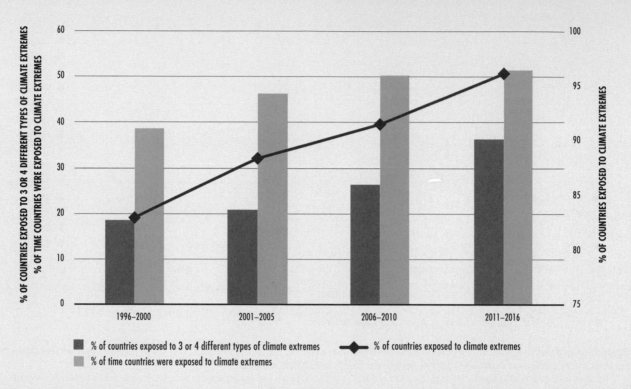

Legend:
- % of countries exposed to 3 or 4 different types of climate extremes
- % of time countries were exposed to climate extremes
- % of countries exposed to climate extremes

NOTES: Percentage of low- and middle-income countries exposed to three or four types of climate extremes (extreme heat, drought, floods and storms) during any of the periods shown; percentage of time (based on the average number of years within a period) that a country was exposed to climate extremes; and percentage of countries exposed to at least one climate extreme in each period. Results are presented using five-year periods, except for 2011–2016 which is a six-year period. See Annex 2 for definition and methodology. Analysis is only for low- and middle-income countries.
SOURCE: C. Holleman, F. Rembold and O. Crespo (forthcoming). *The impact of climate variability and extremes on agriculture and food security: an analysis of the evidence and case studies.* FAO Agricultural Development Economics Technical Study 4. Rome, FAO.

early or delayed onset of seasons, 29 experienced seasons of shorter length, and 28 experienced both. This is an added risk factor affecting food security and nutrition. Furthermore we observe that all countries exposed to interseasonal variability are also exposed to climate extremes.

Undernourishment has been on the rise over the past three years and, as explained here, exposure of countries to climate variability and extremes is also a rising trend. Nonetheless, the latter seems to have started much earlier than the former in low- and middle-income countries. This begs the question: Are these trends associated? It would appear so.

Simple correlations show higher levels of food insecurity in countries with high levels of exposure to climate shocks.[89] Those countries experiencing climate extremes for more than three years in the period of 2011–2016 are defined as having high exposure, irrespective of whether they are countries of low or middle income. This indicates a high frequency of exposure to climate shocks, repeated within a short period of time.

In 2017, the average of the PoU in countries with high exposure to climate shocks was 3.2 percentage points above that of countries with low or no exposure (Figure 25). Even more striking is that countries with high exposure have more than

FIGURE 25
HIGHER PREVALENCE AND NUMBER OF UNDERNOURISHED PEOPLE IN COUNTRIES WITH HIGH EXPOSURE TO CLIMATE EXTREMES

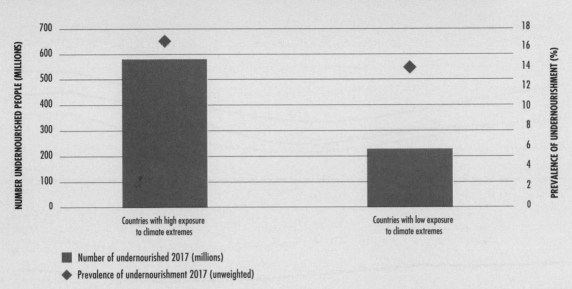

■ Number of undernourished 2017 (millions)
◆ Prevalence of undernourishment 2017 (unweighted)

NOTES: Prevalence (unweighted) and number of undernourished people in low- and middle-income countries with high and low exposure to climate extremes during the period of 2011–2016. Countries with high exposure are defined as being exposed to climate extremes (heat, drought, floods and storms) for more than 66 percent of the time, i.e. for more than three years in the period 2011–2016; low exposure is three years or less. See Annex 2 for the list of countries with high exposure to climate extremes and methodology.
SOURCE: C. Holleman, F. Rembold and O. Crespo (forthcoming). *The impact of climate variability and extremes on agriculture and food security: an analysis of the evidence and case studies.* FAO Agricultural Development Economics Technical Study 4. Rome, FAO, for classification of countries with high and low exposure to climate extremes; FAO for data on prevalence of undernourishment.

double the number of undernourished people (351 million more) as those without high exposure.

Of the 51 countries identified as experiencing high exposure to climate extremes in 2011–2016, 23.5 percent are low-income countries and 76.5 percent are middle-income. In terms of geographical location, most (76 percent) are in Africa and Asia (39 and 37 percent, respectively), 15.5 percent in Latin America and the Caribbean, and the rest in Oceania and Europe (see Annex 2).

Because low- and middle-income countries are increasingly exposed to climate extremes, the vulnerability to these events is an important risk factor for food security and nutrition that merits more study. Vulnerability here refers to the conditions that raise the probability that climate extremes will negatively affect food security. Vulnerability of national agriculture production and yields to climate extremes, along with increased vulnerability of related food supply

chains and natural resource-based livelihoods, need to be at the centre of the analysis.

There are marked (i.e. statistically significant) differences in the PoU of the 128 countries identified in the present analysis when considering high levels of vulnerability to climate extremes that pose risks to food security. Risks exist where cereal production and/or yields are sensitive to climate variability and extremes, and where livelihoods are sensitive to climate. Risks also exist where severe drought conditions correspond to rises in the PoU (see Box 9 for definitions and indicators analysed).

For example, analysis conducted for this report shows that, in 2017, the average of the PoU was 15.4 percent for all countries exposed to climate extremes. At the same time, the PoU was 20 percent for countries that additionally show high vulnerability of agriculture production/ yields to climate variability, or 22.4 percent for

FIGURE 26
UNDERNOURISHMENT IS HIGHER WHEN EXPOSURE TO CLIMATE EXTREMES IS COMPOUNDED BY HIGH LEVELS OF VULNERABILITY IN AGRICULTURE

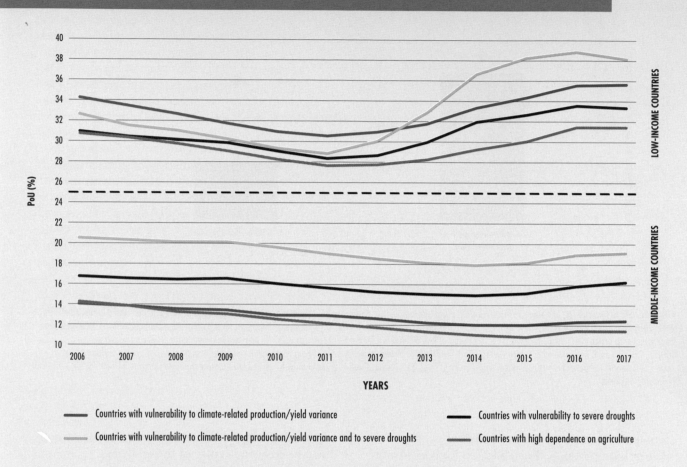

NOTES: The estimates in the graph refer to the unweighted population average of the PoU in a sample of 128 low- and middle-income countries with exposure to climate extremes, for countries with different high vulnerabilities as identified in Box 9. Exposure to climate extremes is not differentiated in this figure, i.e. it includes all levels of exposure to climate extremes, both high and low exposure. See Annex 2 for more detailed definitions and methodology of the different types of vulnerability to climate variability and extremes.
SOURCE: C. Holleman, F. Rembold and O. Crespo (forthcoming). *The impact of climate variability and extremes on agriculture and food security: an analysis of the evidence and case studies.* FAO Agricultural Development Economics Technical Study 4. Rome, FAO, for exposure (both low and high) to climate extremes; FAO for data on prevalence of undernourishment.

countries with high PoU vulnerability to severe drought. When there is both high vulnerability of agriculture production/yields and high PoU sensitivity to severe drought, the PoU is 9.8 points higher (25.2 percent). A high dependence on agriculture, as measured by the number of people employed in the sector, leaves the PoU 9.6 percentage points higher (25 percent); for low-income countries, the increase is equal to 13.6 percentage points (29 percent).

What is striking is that the uptick in PoU occurs earlier in time for low-income countries, and with sharper increases, especially in those with high vulnerability of agriculture production/yields and high sensitivity of PoU to severe drought (Figure 26).[90]

The finding is different for middle-income countries where the rise in PoU is less pronounced and occurs later (from 2015–2016). Here the increase in PoU is also more marked for

BOX 9
FOOD SECURITY VULNERABILITY FACTORS ANALYSED

Vulnerability refers to the conditions that increase the probability that climate extremes will negatively affect food security. Although there are many other vulnerability factors, the below are selected due to their relative importance for food availability and access as identified later in the report.

▶ *Vulnerability related to climate-sensitive production and/or yields*: countries with at least part of their national cereal production or yield variance explained by climate factors – i.e. there is a high and statistically significant association between production and climate or biophysical indicators such as temperature, rainfall and vegetation growth (see Figure 29a for production and see report cited below in source for yield).

▶ *Vulnerability related to severe drought food security sensitivity*: countries with severe drought warnings corresponding with the occurrence of PoU change points (see Figure 23).

▶ *Vulnerability related to high dependence on agriculture*: countries with a high dependence on agriculture (measured by the percentage of people employed in the sector according to World Bank, 2017), where it is expected that many derive their livelihood and income from the sector.

SOURCE: C. Holleman, F. Rembold and O. Crespo (forthcoming). *The impact of climate variability and extremes on agriculture and food security: an analysis of the evidence and case studies.* FAO Agricultural Development Economics Technical Study 4. Rome, FAO. See Annex 2 for full definitions and methodology.

countries with high agriculture production/yield vulnerability and high vulnerability to severe drought. This tends to indicate that middle-income countries were able to absorb the impacts of increased exposure to climate extremes, but may not have been able to cope as well in the 2015–2016 period, possibly due to the severity of exposure to El Niño. Other factors may have also come into play during this period, for example the economic slowdowns that many

Latin American countries experienced, which reduced the fiscal space to implement social programmes and thus diminished these countries' capacity to cope with the aftermath of extreme climate events.

The increase in PoU is even more pronounced and begins in 2011 for those countries with both high exposure to climate extremes (more than 66 percent of the time) and high levels of vulnerability (Figure 27).

Countries highly dependent on agriculture show the highest levels of PoU, whereas countries experiencing both climate-sensitive vulnerability of production/yields and vulnerability to severe drought show the sharpest increase in undernourishment starting from 2011, followed by countries with either production/yield vulnerability or vulnerability to severe drought.

What is striking about Figure 27 is that, as noted above, most countries (close to three-quarters) with high exposure to climate extremes are actually middle-income countries, yet we see an uptick in PoU from 2011 (Figure 26) which is mostly driven by low-income countries.

Climate extremes as a major driver of global food crises
In 2017, almost 124 million people across 51 countries and territories faced "crisis" levels of acute food insecurity or worse (IPC Phase 3 and above or equivalent)[91] requiring urgent humanitarian assistance to safeguard their lives and preserve their livelihoods. In 34 of these countries more than 76 percent of the total populations facing crisis levels of acute food insecurity or worse – nearly 95 million people – were also affected by climate shocks and extremes (Table 7).

Where conflict and climate shocks occur together, the impact on acute food insecurity is more severe. In 2017, 14 out of the 34 food-crisis countries experienced the double impact of both conflict and climate shocks, which led to significant increases in the severity of acute food insecurity. A total of 65.8 million people (IPC Phase 3 and above) required immediate humanitarian assistance in 2017, of which 15.5 million people suffered very extreme levels

FIGURE 27
UNDERNOURISHMENT IS HIGHER FOR COUNTRIES WITH BOTH HIGH EXPOSURE TO CLIMATE EXTREMES AND HIGH VULNERABILITY

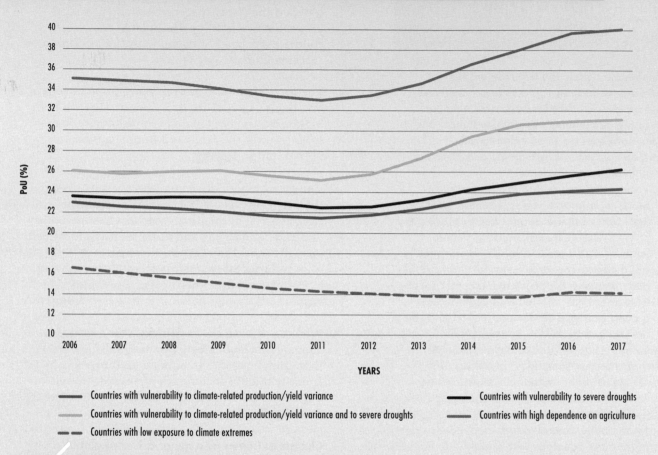

Countries with vulnerability to climate-related production/yield variance

Countries with vulnerability to climate-related production/yield variance and to severe droughts

Countries with low exposure to climate extremes

Countries with vulnerability to severe droughts

Countries with high dependence on agriculture

NOTES: Low- and middle-income countries with high exposure are defined as exposed to climate extremes (heat, drought, floods and storms) for more than 66 percent of the time, i.e. more than three years in the period 2011–2016. The estimates in the figure refer to unweighted population average of the prevalence of undernourishment in a sample of 51 low- and middle-income countries with high exposure to climate extremes in 2011–2016, for countries showing different combinations of vulnerabilities identified in Box 9 and for 77 low- and middle-income countries with low exposure to climate extremes. See Annex 2 for more detailed definitions and methodology of the different types of vulnerability to climate variability and extremes.
SOURCE: C. Holleman, F. Rembold and O. Crespo (forthcoming). *The impact of climate variability and extremes on agriculture and food security: an analysis of the evidence and case studies*. FAO Agricultural Development Economics Technical Study 4. Rome, FAO, for exposure (both low and high) to climate extremes; FAO for data on prevalence of undernourishment.

of acute food insecurity requiring urgent life-saving assistance (IPC 4 and above).

Most climate-related food crisis countries are not affected by conflict, yet climate shocks and stressors are a major factor driving emergency levels of acute food insecurity (20 out of 34 countries). For these climate-affected food crisis countries, 29 million people required humanitarian assistance (IPC Phase 3 and above), including 3.9 million people in need of urgent life-saving emergency assistance (IPC 4 and above).

Drought is a driving climate factor in 21 out of the 34 countries. However, drought occurs without other climate shocks in only seven of these countries. In most cases, countries are also exposed to drought combined with floods, cyclones, and other less extreme but equally detrimental climate events, including dry spells and erratic rainfall, and late onset of rainy seasons (Table 7).

Africa is the region where climate shocks and stressors had the biggest impact on acute food insecurity and malnutrition, affecting 59 million people in 24 countries and requiring urgent humanitarian action. »

TABLE 7
CLIMATE SHOCKS WERE ONE OF THE LEADING CAUSES OF FOOD CRISIS SITUATIONS IN 2017

Regions	Climate shocks		Countries affected by climate shocks (also affected by conflict ✱)	Number of people (millions)	
				IPC/CH Phase 3 (Crisis)	IPC/CH Phase 4 (Emergency)
Africa	Droughts		✱ Burundi, Djibouti, Eswatini, Kenya, Lesotho, Namibia, ✱ Somalia	8.4	2.3
	Dry spells/low rainfall		Angola, ✱ Chad, ✱ South Sudan, Uganda	6.9	1.7
	Seasonal variability (late onset of the rainy season)		✱ Sudan, Zambia	3.7	0.1
	Late onset and dry spells/erratic rainfalls		✱ Cameroon, Gambia, Mauritania (early cessation rainy season), Niger, United Republic of Tanzania	5.7	0.1
	Late onset and floods		✱ Guinea-Bissau	0.3	0
	Droughts and other climate shocks		Malawi	5.1	N/A
			✱ Ethiopia	8.5	N/A
			Zimbabwe	3.5	0.6
			✱ Democratic Republic of the Congo	6.2	1.5
			Madagascar, Mozambique	3.4	1.3
Asia	Floods and other climate shocks		✱ Afghanistan, ✱ Nepal, ✱ Pakistan	7.8	3.3
			Bangladesh	2.9	0.5
			✱ Sri Lanka, ✱ Yemen	11.1	6.8
Latin America and the Caribbean	Drought and other climate shocks		Guatemala, Haiti	2.1	0.7
			Honduras	0.4	0
				76.0	18.9
				94.9	

✱ Countries affected by conflicts   Countries affected by dry spells   Countries affected by seasonal variability   Countries affected by floods

Countries affected by droughts   Countries affected by flash flood   Countries affected by storms

NOTES: This table is elaborated on the basis of the Global Food Crisis Report (GFCR 2018). The table reports the number of people who are food insecure classified according to the Integrated Food Security Phase Classification (IPC) or the *Cadre Harmonisé* (CH) and reports on the occurrence of specific climate shocks (droughts, floods and cyclones) which are drivers contributing to food insecurity. This information is complemented with information on other types of climate shocks linked with food insecurity (dry spells, flash floods and seasonal variability). Information for these were identified from the GFCR 2018 and the FAO Global Information and Early Warning System for Food Security and Agriculture (GIEWS) Country briefs. Population in IPC Phase 4 for South Sudan also includes population in IPC Phase 5. Some countries are not included in the report due to lack of recently validated data or because variations in the geographical coverage of IPC or CH analysis represent a technical limitation in showing trends for certain countries.
SOURCE: FAO elaboration based on FSIN. 2018. *Global Report on Food Crisis 2017.*

» Areas where climate shocks and conflict interact to drive food crises have very high to high prevalence rates of acute malnutrition in children under five years of age – these include Darfur in the Sudan (28 percent), South Sudan (23 percent), the Lake region of Chad (18 percent), Yemen (10–15 percent), the Diffa region of the Niger (11 percent), the Democratic Republic of the Congo (8–10 percent), and Afghanistan (9.5 percent).

There is also a high burden of acute malnutrition in areas or countries affected by drought/floods, including northern Kenya, the Sindh province in Pakistan, Ethiopia and Madagascar.[92] Climate shocks exacerbate the factors that underlie acute malnutrition, including: high levels of food insecurity; inadequate access to diverse and nutrient-rich foods; high prevalence of diseases, such as diarrhoea, malaria and fever; poor access to primary health care and safe water; inadequate sanitation; and suboptimal breastfeeding practices.

Many studies have shown that the health and nutritional status of children in particular is especially vulnerable to climate-related disasters, both in the emergency phase and – due to malnutrition and undernutrition – also in the aftermath.[93] The impacts of floods and droughts on peaks in acute malnutrition (through crop damage or disease) are well documented.[94]

Summary

In the twenty years (1996–2016) considered in the analysis presented here, both the frequency and intensity of countries' exposure to climate extremes have increased. As a result, more countries are vulnerable to the risk of food insecurity and malnutrition. Where agriculture production, food systems and livelihoods are vulnerable to climate variability and extremes, countries face the greatest risk of food insecurity and malnutrition.

Although climate variability and extremes are not the only factor driving the observed increases in global hunger, the analysis indicates that they are important for some countries. They also exacerbate other driving factors of food insecurity and malnutrition, such as conflict, economic slowdowns and poverty.[95] It is thus critical to investigate in more detail how climate variability and extremes can undermine the different dimensions of food security (food availability, access, utilization and stability) and nutrition. ∎

2.2 HOW DO CHANGING CLIMATE VARIABILITY AND EXTREMES AFFECT THE IMMEDIATE AND UNDERLYING CAUSES OF FOOD INSECURITY AND MALNUTRITION?

KEY MESSAGES

➜ Climate variability and extremes are undermining in multiple ways food availability, access, utilization and stability, as well as feeding, caregiving and health practices.

➜ Direct and indirect climate-driven impacts have a cumulative effect, leading to a downward spiral of increased food insecurity and malnutrition.

➜ Climate variability and extremes are harming agricultural productivity, food production and cropping patterns, thus contributing to food availability shortfalls.

➜ Food price spikes and volatility, often combined with losses in agricultural income, follow climate extremes, reducing food access and negatively affecting the quantity, quality and dietary diversity of food consumed.

➜ Changes in climate impact heavily on nutrition through: impaired nutrient quality and dietary diversity of foods produced and consumed; effects on water and sanitation, with their implications for patterns of health risks and disease; and changes in maternal and child care and breastfeeding.

To adequately respond to the challenges that changing climate variability and extremes create for food security and nutrition, it is critical to factor in the multiple direct and indirect impacts

FIGURE 28
LINKS BETWEEN FOOD SECURITY AND NUTRITION, AND THE UNDERLYING CAUSES OF FOOD INSECURITY AND MALNUTRITION

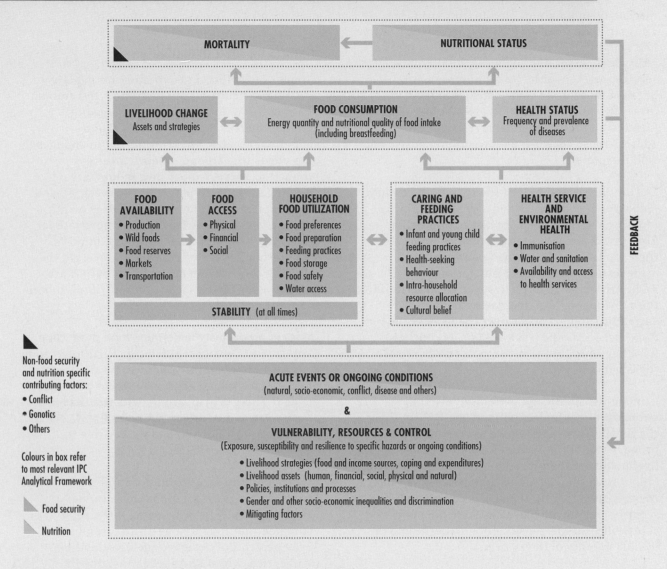

SOURCE: Integrated Food Security Phase Classification (IPC) (forthcoming). *IPC Technical Manual 3.0.*

that occur in different combinations and of varying durations. Climate impacts flow through different channels, exacerbating the basic causes of food insecurity and malnutrition.

For example, a direct impact occurs when drought undermines crop yields, which then results in reduced food production. On the other hand, crop failures can indirectly reduce food access if food prices rise significantly. Similarly, floods that reduce access to safe water and adequate sanitation can indirectly affect the utilization of

food and nutrition, as a result of reduced quality and safety of food and disease outbreaks. The cumulative effect of these direct and indirect impacts leads to a downward spiral of increased food insecurity and malnutrition.

Figure 28 presents a conceptual framework that shows links between food security and nutrition and the basic and underlying factors that drive food security and nutritional status. It shows how, whether acute or ongoing, climate variability and extremes can influence the immediate and underlying causes of food insecurity and malnutrition in all their forms. These include food availability, access, utilization and stability (see Annex 4 Glossary), as well as individual caregiving practices, quality health services and a healthy living environment. Because these basic causal factors of food insecurity and malnutrition are all potentially affected and interdependent, responses to address these causes must be comprehensive and well integrated.

During the second half of the twentieth century, global food availability and access developed rapidly enough to keep abreast of population growth. As a result, many countries improved their food security and made impressive achievements in reducing hunger and malnutrition by 2015.[96] However, as described in the previous section, increasing climate variability and extremes over the last decade (together with other factors such as conflicts) have begun to threaten and potentially reverse these gains.[97]

Climate variability and extremes have the strongest direct impact on food availability, given the sensitivity of agriculture to climate and the primary role of the sector as a source of food and livelihoods for the rural poor. However, the overall fallout is far more complex and greater than the impacts on agricultural productivity alone.[98] Food security and nutrition are also dependent on food access, utilization, consumption patterns and the overall stability of the system.

Nutritional status is determined by the interaction between dietary intake and health status. Illness and disease become more likely if climate variability and extremes prompt people to consume inadequate or insufficient food, or to engage in crisis and emergency coping strategies. There can be further repercussions for access to and utilization of food if people's immune systems are compromised or if people are more exposed to disease risk factors vectors, particularly in situations with insufficient health services.

Unravelling how climate variability and extremes are negatively affecting food security and nutrition is an important first step towards designing effective strategies, policies and programmes to reverse these impacts.

Impacts on food availability

Climate variability and extremes are negatively affecting agricultural productivity – the amount of agricultural outputs per inputs used to generate them – at global, national and subnational scales. This is reflected in changes in crop yields (the amount of agricultural production harvested per unit of land area), cropping areas (area planted or harvested), and cropping intensity (number of crops grown within a year). Countries try to compensate for domestic production losses through imports, though supplies are often limited. Overall, the resulting shortfalls in agricultural output are damaging for food security and nutrition in both the short and long term.

Losses in productivity undermine food production

Crop yields in many countries have suffered from changes in temperature and precipitation, which have affected global aggregate wheat and maize yields.[99] There is also strong evidence that climate variability driven by major ENSO events associated with El Niño plays a key role in decreasing crop yields.[100]

Studies point to significant heat and water stress resulting in significant global interannual variability of yields for wheat and maize.[101] It is estimated that roughly one-third (around 32–39 percent) of observed yield variability (maize, rice, wheat and soybean) is due to climate factors.[102]

Throughout the growing season, crops are highly sensitive to extreme daytime temperatures of around 30 °C, resulting in lower yields.[103] Analysis of global crop yield variability during the 1961–2014 period shows that heat and dryness significantly reduced yields of maize, soybeans and wheat, although the effects for rice were not significant.[104]

Most regions, particularly those with large numbers of undernourished people, are experiencing reduced yields due to increased climate variability and extremes. In sub-Saharan Africa, a region that already has the lowest crop yields globally, increasing temperatures reduced yields for maize, sorghum and groundnuts.[105] In rural India, higher numbers of hot days during the agricultural season are leading to lower crop yields.[106] There are regions that show increased yields due to changes in climate, but these are fewer: for example, north-east China, the United Kingdom and Ireland have seen some improvement in yields, given their higher latitudes.[107]

While the impact of drought on decreasing crop yields is widely documented,[108] the effects of other climate extremes, such as tropical cyclones, are not well quantified, though their influence in some regions is evident. Crop destruction due to tropical cyclones can include salt damage from tides blowing inland, insufficient oxygen caused by overhead flooding, flash floods, wind damage to plants, and water stress induced by enforced respiration, all of which can occur at the same time.[109] For example, in Bangladesh cyclones cause increased salinity from seawater to coastal and freshwater fishery communities, negatively affecting production due to insufficient access to fresh water.[110]

A focus exclusively on yields may bias assessments of the vulnerability of agriculture to climate shocks. Although there is no global overview, a number of case studies provide evidence that both cropping intensity and planted areas are negatively affected by climate variations and extremes.

For example, in the Viet Nam Mekong Delta, variations in the timing and extent of flooding in the wet season and salinity intrusion in the dry season are affecting rice cropping cycles.[111] Severe floods in 2000 led to crop failure, except for floating rice varieties. In contrast, below-normal seasonal rainfall in 2004 reduced water availability for irrigation due to high salinity, and as a result the dry-season rice that year could not be harvested.[112] Based on the existing country evidence, it is clear that efforts to reduce climate impacts on agriculture should seek to limit production losses resulting not only from crop yields, but also from changes in cropping area and frequency.[113]

Of course, climate impacts vary between regions, countries, and within a given

country. Differences in overall aggregate impacts on national food production arise not only due to variations in type and geographical distribution of climate variability and extremes, but also due to the diversity and complexity of agricultural systems. Divergences exist between crops, cropping patterns, farming technologies (e.g. rainfed vs irrigated, high and low input ratios, nomadic pastoral vs intensive livestock production) and agriculture management systems.

Despite these nuanced and varied elements, there is evidence that for many countries, climate factors at least partially explain national cereal production variance (Figure 29a). Especially in semi-arid climate regions such as Central Asia, the Near East, and Northern Africa, cereal production is highly dependent on climate variability. In these regions, it is not unusual to have 80 percent or more of interannual production variability explained by climate.

Although the influence of climate on production can be seen in a large number of countries, the relationship is strongest but also most complex in Africa. In this continent, the production of each country shows a different mix of climate variable dependence, both in terms of strength and correlation. In contrast, in many Asian countries – such as China, India and Kazakhstan – there is no significant correlation with single climate indicators, but only with biophysical indicators such as NDVI, partly due to the complex dependence of agricultural vegetation growth on many climate and non-climate factors.

Drought is one of the most important climate events that have been shown to have a negative impact on production. For many countries, there is a high negative correlation between drought indicators and food production (Figure 29b). The highest correlations occur in semi-arid countries or drought-prone continental climates (e.g. Central Asia), while in many equatorial areas there is no correlation between drought indicators and production (e.g. central Africa, Central America).

FIGURE 29

EFFECT OF CLIMATE VARIABILITY AND DROUGHT ON NATIONAL CEREAL PRODUCTION OF LOW- AND MIDDLE-INCOME COUNTRIES, 2001–2017

A) RELATIONSHIP BETWEEN NATIONAL CEREAL PRODUCTION AND TEMPERATURE, RAINFALL AND VEGETATIVE GROWTH

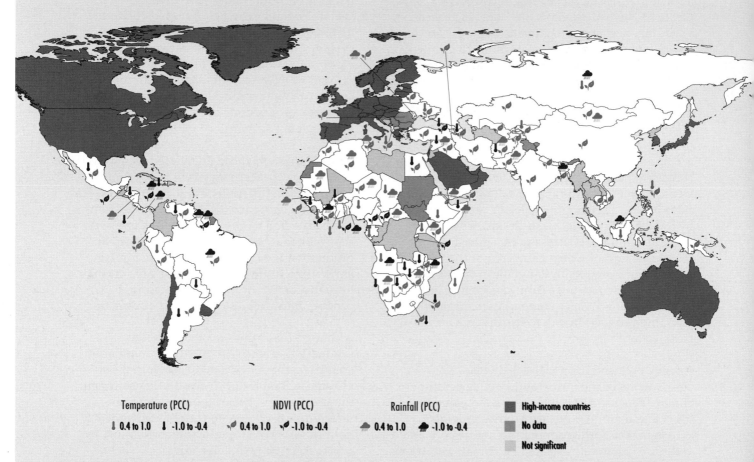

Temperature (PCC)	NDVI (PCC)	Rainfall (PCC)	
0.4 to 1.0 -1.0 to -0.4	0.4 to 1.0 -1.0 to -0.4	0.4 to 1.0 -1.0 to -0.4	High-income countries
			No data
			Not significant

NOTES: Figure shows where part of cereal production variability in low- and middle-income countries is explained by a) mean annual temperature, cumulative NDVI over the growing season and cumulative annual rainfall, and b) two climate indicators that measure drought: Anomaly Hotspots of Agriculture Production (ASAP) and Agriculture Stress Index System (ASIS). Colours of the symbols reflect the sign of the correlation (green = positive, red = negative), as provided by the Pearson correlation coefficient (PCC). See Annex 3 for data sources and methodology. The final boundary between the Republic of the Sudan and the Republic of South Sudan has not yet been determined. Final status of the Abyei area has not yet been determined.

Climate variability and extremes may not always affect aggregate national food production but can significantly affect subnational areas with often devastating impacts on the food security and nutrition situation of their populations. This is especially the case in areas dominated by small-scale family farmers and pastoralists, whose production losses may be significant for their own livelihoods and food security and nutrition situation, but not necessarily for national food production.

For example, Ethiopia has experienced large increases in national cereal production in recent decades, yet it regularly reports acute and localized food insecurity and malnutrition crises, often associated with droughts.[114] The greatest adverse impacts occur in the most marginal livelihood zones in the drier east of the country. Drought incidences are usually relatively local, with serious impacts on local production and livelihoods that leave people unable to meet their food needs by buying from other regions, even though, on the

B) RELATIONSHIP BETWEEN NATIONAL CEREAL PRODUCTION AND MEASURES OF DROUGHT (ASAP AND ASIS)

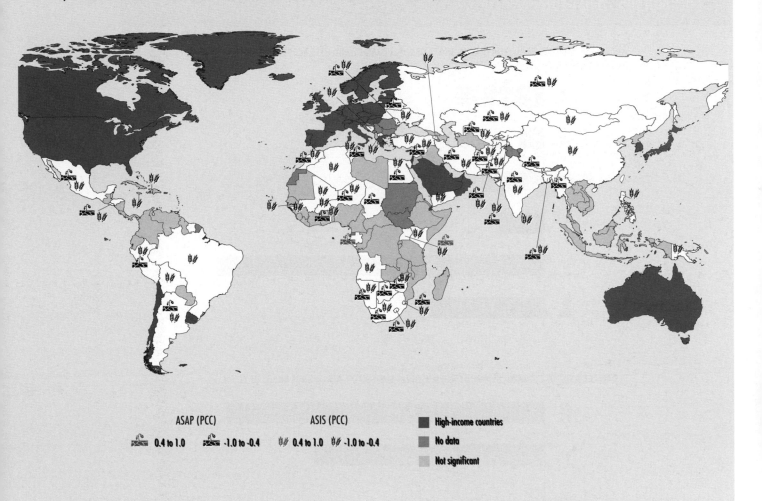

ASAP (PCC)
0.4 to 1.0   -1.0 to -0.4

ASIS (PCC)
0.4 to 1.0   -1.0 to -0.4

High-income countries
No data
Not significant

SOURCE: C. Holleman, F. Rembold and O. Crespo (forthcoming). *The impact of climate variability and extremes on agriculture and food security: an analysis of the evidence and case studies.* FAO Agricultural Development Economics Technical Study 4. Rome, FAO. Data sources are EC-JRC for ASAP and FAO for ASIS.

whole, the country is no worse off than in any other year.[115]

Other examples include the Afram Plains region of Ghana, where farmers report delays in the onset of the rainy season, mid-season heatwaves, and high-intensity rains that cause flooding, resulting in crop loss and low yields that reduce the availability of household food. However, due to the localized and marginal nature of most of the region's agriculture areas, this is not reflected in national production losses.[116] Similarly, a study

in China found that even though the most negative impacts of precipitation for each province during 1980–2008 occurred simultaneously, these did not lead to a serious reduction of crop harvests at the national level.[117]

The focus on drought is well justified – 83 percent of the damage and losses caused by droughts affect the agricultural sector, especially crop production and livestock (Figure 30).[118] Fisheries and forestry show lower levels of damage and losses, but they can be significant

FIGURE 30
CROP AND LIVESTOCK SUB-SECTORS INCUR THE HIGHEST DAMAGES AND LOSSES IN AGRICULTURE DUE TO CLIMATE-RELATED DISASTERS, OF WHICH DROUGHT IS THE MOST DESTRUCTIVE, 2006–2016

A) DAMAGE AND LOSS IN AGRICULTURE AS SHARE OF TOTAL DAMAGE AND LOSS ACROSS ALL SECTORS BY TYPE OF HAZARD

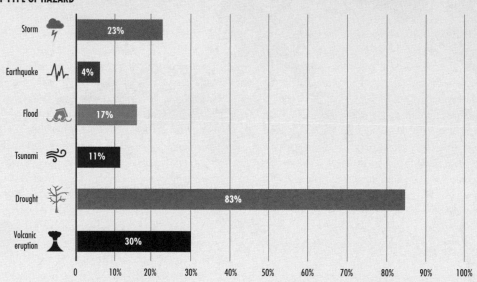

B) DAMAGE AND LOSS IN AGRICULTURE BY AGRICULTURAL SUB-SECTOR, PERCENTAGE SHARE OF TOTAL

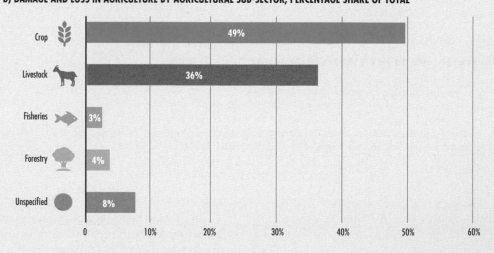

NOTES: FAO, based on Post Disaster Needs Assessments (PDNA), 2006–2016. The sectors of fisheries, aquaculture and forestry often are under-reported. Impact of disasters on forestry is generally acknowledged in assessments, although rarely quantified in monetary terms.
SOURCE: FAO. 2018. *The impact of disasters and crises on agriculture and food security 2017.* Rome.

for populations dependent on these subsectors for their livelihoods and food. Fisheries, an important source of food production for many countries, are most affected by tsunamis and storms. Studies have shown that climate variability affects fisheries directly, as fish populations and fisheries activities are closely linked to weather and climate dynamics. The strongest economic impacts on forestry are caused by floods and storms.[119] While the impact of floods and storms on forests can be significant, deforestation exacerbates the negative impact of floods and storms, triggering a vicious downward cycle.

FAO agricultural databases were used to conduct a statistical analysis of 140 medium- and large-scale climate-related disasters (affecting at least 250 000 people) that occurred in 67 developing countries between 2003 and 2013.[120] The analysis estimates losses equivalent to 333 million tonnes of cereal, pulses, meat, milk and other commodities, or an average of 7 percent of national per capita dietary energy supply (DES) after each natural disaster. This is already significant at the national level, but is likely higher at the subnational one, where losses in calories may increase household food insecurity unless relevant measures are taken to compensate and fill the gap in DES.[121]

The Dry Corridor in Central America – in particular in El Salvador, Guatemala and Honduras – was one of the regions heavily impacted by El Niño in 2015–2016. The drought impact was severe and prolonged, with late and irregular onset of rains, below-average rainfall, above-average temperatures and river levels 20 to 60 percent lower than normal. The drought was one of the worst in the last ten years and resulted in significant reductions in agriculture production, with losses estimated at 50–90 percent of crop harvest.[122] In Guatemala alone, the Ministry of Agriculture, Livestock and Food estimated that 82 000 tonnes of maize were lost, representing a total financial loss of USD 30.8 million, while 118 200 tonnes of black beans were lost, at a cost of USD 102.3 million. More than 3.6 million people were in need of humanitarian assistance as result of this drought.

The same period saw the worst drought in 35 years hit southern Africa, leading to extensive regional-scale crop failure and a regional cereal deficit of 7.9 million tonnes in early 2016.[123] The impacts were magnified further as depleted food supplies and reserves spurred rising food prices. In response, six countries (Botswana, Eswatini,

Lesotho, Malawi, Namibi and Zimbabwe) declared national drought emergencies, while two declared partial drought disasters (Mozambique and South Africa).

At the regional level, the Southern Africa Development Community (SADC) declared a regional drought disaster and issued a regional humanitarian appeal, seeking local and international assistance to cover a response plan funding gap of USD 2.5 billion for an estimated 41 million affected people (about 14 percent of the total SADC population), with 26 million requiring immediate humanitarian assistance.[124]

Production shortfalls lead to increased food imports
Climate variability and extremes also affect food imports as countries try to compensate for domestic production losses.[125] It is expected that, as production falls, exports will follow suit, resulting in a deterioration of trade flows. For low- and middle-income countries, high temperatures, low rainfall and low NDVI generally show a significant correlation with high cereal imports, indicating vulnerability to climate variability and extremes (Figure 31). This applies to the Middle East/North Africa (MENA) countries and those in western and southern Africa, while in eastern Africa and Central America temperature seems to be the single indicator most directly linked to imports.

Nonetheless, as shown in Figure 32, estimated agricultural commodity decreases in exports and increases in imports owing to the harmful effects of climate-related shocks on domestic production tend to be, on aggregate, largest for Asia and Latin America and the Caribbean. This can be considered an indirect effect of losses to domestic production and consequent rise in demand for imported food. In the case of Africa, although the ratio of exports to imports has continuously been falling since the 1970s and the continent became a net food importer in 2000, the findings show that increases in agricultural imports after disasters are proportionally lower than losses in domestic production.[126] In some cases, the compensating increase in imports in Africa can be as much as half the losses. Humanitarian response in Africa is high and can fill some of the gap, but there are still negative consequences for food availability.

FIGURE 31
CLIMATE VARIABILITY AND EXTREMES ARE CORRELATED WITH CEREAL IMPORTS IN MANY LOW- AND MIDDLE-INCOME COUNTRIES

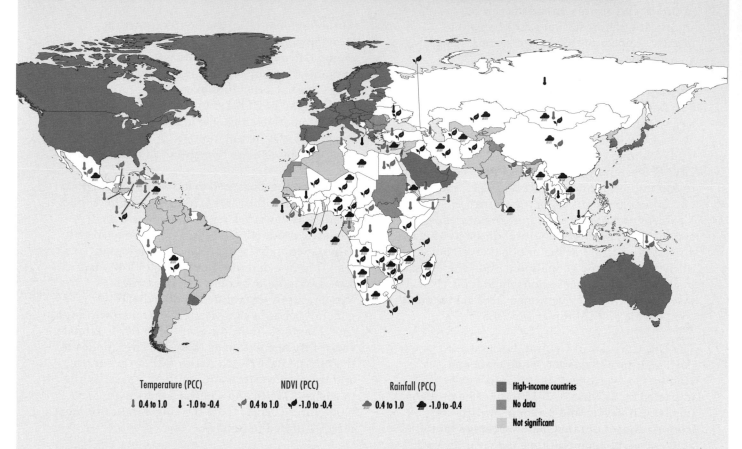

Temperature (PCC)	NDVI (PCC)	Rainfall (PCC)	
↓ 0.4 to 1.0 ↓ -1.0 to -0.4	0.4 to 1.0 -1.0 to -0.4	0.4 to 1.0 -1.0 to -0.4	High-income countries
			No data
			Not significant

NOTES: For low- and middle-income countries, showing where part of cereal import variability is explained by climate indicators. Colours of the symbols reflect the sign of the correlation (green = positive, red = negative), as provided by the Pearson correlation coefficient (PCC). Shows correlation results between total annual cereal imports (data source: FAO GIEWS) with cumulative precipitation or rainfall, annual average temperature and cumulative NDVI (normalized difference vegetation index) during active crop season. All climate indicators extracted over agriculture cropping areas. See Annex 3 for data sources and methodology. The final boundary between the Republic of the Sudan and the Republic of South Sudan has not yet been determined. Final status of the Abyei area has not yet been determined.
SOURCE: C. Holleman, F. Rembold and O. Crespo (forthcoming). *The impact of climate variability and extremes on agriculture and food security: an analysis of the evidence and case studies.* FAO Agricultural Development Economics Technical Study 4. Rome, FAO.

An in-depth analysis of the impact of drought in sub-Saharan Africa provides a stark illustration of this. The study estimates that after the occurrence of droughts between 1991 and 2011 in the region, food imports increased by USD 6 billion and exports of the same commodities fell by nearly USD 2 billion. Further, countries lost an average of 3.5 percent of agriculture value-added growth after each drought – a figure that is likely to be more acute at the subnational level.[127]

Medium- and long-term impacts on food availability
Beyond production losses and trade deterioration, medium- and large-scale disasters can lead to significant impacts across the food value chain, with negative consequences on sector growth, food and non-food agro-industries and ultimately national economies. In fact, these disasters can inflict high levels of damage and economic loss on agriculture (Figure 33). The financial cost to developing countries alone, in terms of losses to crops and livestock, was estimated at USD

FIGURE 32
INCREASES IN IMPORTS AND DECREASES IN EXPORTS OF AGRICULTURAL COMMODITIES AFTER CLIMATE-RELATED DISASTERS BY REGION, 2003–2011

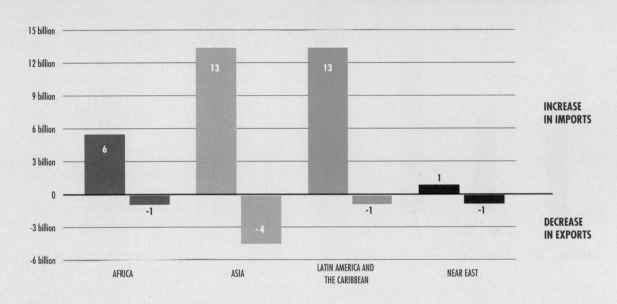

NOTE: Increases in imports and decreases in exports of agriculture commodities in USD by region.
SOURCE: FAO. 2015. *Impact of disaster on agriculture and food security*. Rome.

96 billion for the decade 2005–2015.[128] For many countries, it can take years to recover from damages and losses and the negative effects can extend to the long term where agriculture growth and lifelong nutrition and health (and therefore economic productivity) are affected.

More than 25 percent of all economic losses and damages inflicted by medium- and large-scale climate-induced hazards in developing countries occur in the agriculture sector. Where extreme climate events lead to recurring climate-related disasters, the accumulated costs for the agriculture sector are even more significant. For example, between 2006 and 2013 the Philippines was struck by 75 disasters – mostly typhoons, tropical storms and floods. These caused damages and losses of some USD 3.8 billion to the country's agriculture sector, an average of USD 477 million each year – about one-quarter of the national budget allocated to the sector in 2014.[129]

Pakistan's agriculture sector was affected by three consecutive climate-related disasters (cyclone/floods in 2007; floods in 2010; floods in 2011), which together caused USD 7.6 billion in accumulated damages and losses. This is almost four times what the Government of Pakistan spent on the agriculture sector between 2008 and 2011.[130]

The FAO analysis noted above also showed a significant negative trend in agriculture value-added growth in 55 percent of the disasters.[131] The study found that after each disaster there is an average loss of 2.6 percent of national agriculture value-added growth, with a much more significant impact likely at subnational levels.

This section focuses mainly on the production of primary staple crops, for which data are widely available. However, attention is drawn to the fact that there are other important crops of food production that are relevant for people's dietary

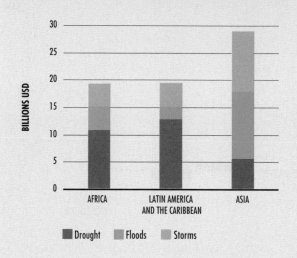

FIGURE 33
CROP AND LIVESTOCK LOSSES CAUSED BY CLIMATE-RELATED DISASTERS BY REGION, 2004–2015

NOTE: Climate-related disasters in the analysis include drought, floods and storms.
SOURCE: FAO. 2015. *Impact of disaster on agriculture and food security*. Rome.

needs and nutrition (fruits and vegetables, legumes other than soybean, etc.) that are not well researched. Future research needs to address the data gap on how climate variability and extremes affect production of these foods.[132]

Impacts on food access

The impacts on production discussed up to this point will inevitably translate into loss of income for people whose livelihoods depend on agriculture and natural resources, reducing their ability to access food. This is another key factor to keep in mind in understanding how climate variability and extremes affect the immediate and underlying causes of food insecurity and malnutrition (Figure 28).

Spikes in food prices and volatility follow climate extremes

Climate anomalies, and in particular extreme events, alter agricultural yields, production and stocks. The critical aspect now is the related effects

on prices. Episodes of high food price volatility pose a major threat to food access, especially in low- and middle-income countries and among poorer groups in high-income countries.

Substantial efforts have been made recently to link the effects of climate on crop yields to prices, income and trade.[133] There is strong statistical evidence that the price of a food basket in communities affected by floods, droughts or cyclones is higher than in control communities – and, interestingly, the effect can last for up to nine months.[134]

Although prices depend on many factors, there is evidence from correlation analyses that higher average temperatures coincide with higher maize prices in some countries, such as Bangladesh, Benin, Eritrea, Ethiopia, Malawi, Nicaragua, Togo and Yemen.[135] The positive temperature and price correlation is visible also for some wheat-producing countries, and the relationship appears typical for wheat produced in tropical countries, for example in Eritrea, Ethiopia, the Sudan and Yemen.[136]

A study covering the period 1960–2014 found evidence that the effects of variability in climate shocks on international maize price volatility intensified during the El Niño phase in spring/summer. Soybean price volatility was also found to respond to climate variability, decreasing slightly during autumn/winter meteorological seasons and increasing during spring/summer.[137]

The impact of price volatility falls heaviest on the urban poor, who may spend as much as 75 percent of their income on food.[138] However, sharp food price increases and price volatility can also significantly undermine the livelihoods and income of small-scale food producers, agriculture labourers and the rural poor who are net food buyers, forcing them to reduce their consumption in quantity and quality.

Global food price spikes often follow climate extremes in major producing countries. Figure 34 shows trends in international food and cereal prices, with vertical lines indicating events when a top five global producer of a crop had yields 25 percent below the trend line, indicating a seasonal climate extreme. In many of these instances, global food prices rose.

FIGURE 34
FOOD PRICE SPIKES FOLLOW CLIMATE EXTREMES FOR TOP GLOBAL CEREAL PRODUCERS, 1990–2016

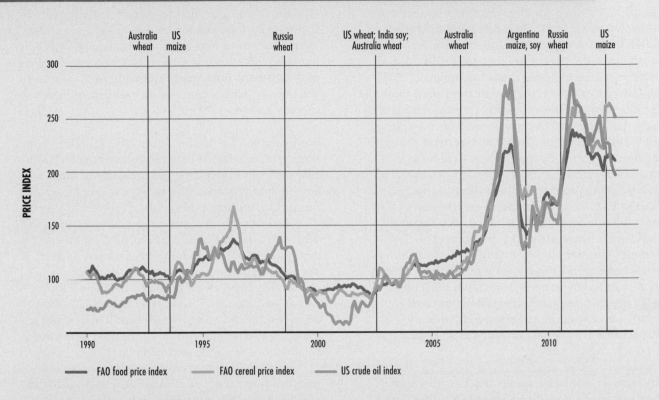

NOTES: The plot shows the history of FAO food and cereal price indices (composite measures of food prices), with vertical lines indicating events when a top five producer of a crop had yields 25 percent below the trend line (indicative of a seasonal climate extreme). All indices are expressed as a percentage of 2002–2004 averages. Food price and crop yield data from FAO (www.fao.org/worldfoodsituation/foodpricesindex and http://faostat.fao.org) and oil price data from U.S. Energy Information Administration (www.eia.gov).
SOURCE: IPCC. 2014. *Climate Change 2014: Impacts, Adaptation, and Vulnerability. Part A: Global and Sectoral Aspects. Contribution of Working Group II to the Fifth Assessment Report of the Intergovernmental Panel on Climate Change* [C.B. Field, V.R. Barros, D.J. Dokken, K.J. Mach, M.D. Mastrandrea, T.E. Bilir, M. Chatterjee, K.L. Ebi, Y.O. Estrada, R.C. Genova, B. Girma, E.S. Kissel, A.N. Levy, S. MacCracken, P.R. Mastrandrea and L.L. White, eds]. Cambridge, UK and New York, USA. Cambridge University Press.

Climate shocks in major global producers contribute to price increases and volatility; however, other factors also play a strong role, thus making attribution difficult. Public and private sector responses to extreme climate events may lead to serious knock-on effects through trade-induced amplification of climate-related food risks that expand across borders. These risks include food price spikes, food safety issues and interactions with conflict and migration, to name but a few. A clear example of a domestic policy response to food price crises is export bans which, in turn, can contribute to more fluctuations.[139] The stability of food prices is also increasingly associated with fluctuations in energy prices.

Income loss for those whose livelihoods depend on agriculture and natural resources
The majority of people most vulnerable to climate shocks and natural hazards are the world's 2.5 billion small-scale farmers, herders, fishers and forest-dependent communities, who derive their food and income from renewable natural resources.[140] Small-scale farmers produce 63 and 69 percent of the food in Kenya and the United Republic of Tanzania, respectively, whereas 70 percent of small family farms are food

producers in Nepal and 85 percent in the Plurinational State of Bolivia.[141]

Climate shocks not only negatively impact on households' own food production but also rural incomes as agricultural production falls. In food-insecure regions, many small family farms both consume their produce and sell it in local markets. This exposes them to climate variations as they have less of their own food production available for consumption and less to sell. Their income is more seriously constrained to maintain a more costly basic consumption,[142] as demonstrated by a wealth of evidence.

Household studies provide evidence that access to food and income of small family agriculture households is negatively impacted by climate variability and extremes. For example, in the United Republic of Tanzania an increase in the variability of rainfall in the past five to ten years is associated with about a 35 percent decrease in total income and increased variability of temperature is associated with a decrease of about 11 percent in daily calorie intake.

In Malawi, the occurrence of a 1 °C increase in temperature (i.e. 1 degree more than the upper confidence interval of the comfort zone) reduces overall consumption per capita by about 20 percent and food calorie intake by almost 40 percent. In Ethiopia and the Niger, both rainfall and maximum temperature variability are shown to negatively affect household income and consumption expenditure. This points towards the absence of capacity to cope or options for income-smoothing behaviour.[143] There is also evidence that climate shocks not only affect the level of income, but affect also the variability of incomes. Household studies for Malawi and Zambia show that increased variation in seasonal rainfall (defined over 30 years) not only decreases the expected incomes but also increases their variance.[144]

Climate shocks that negatively impact agricultural production also negatively impact demand for agricultural labour, thus indirectly affecting access to food and income for rural agriculture labourers. Given the high level of dependency of poor and food-insecure people on agriculture for their incomes, the financial impacts of climate variability can be high.

For those whose livelihoods depend on livestock, climate shocks can lead to significantly depleted income and food resources due to loss of animals, milk production and trade. Furthermore, many livestock diseases are linked to climate variability and extremes, both geographically and temporally, which can lead to significant losses in income and food.[145]

For example, Rift Valley Fever (RVF), endemic in large parts of Africa, is a mosquito-borne viral disease in livestock that has repeatedly caused severe epidemics leading to high levels of morbidity and mortality among affected animals. RVF outbreaks and patterns are closely associated with shifts from El Niño to La Niña. In East Africa, over half of El Niño occurrences have been accompanied by corresponding RVF outbreaks. An RVF outbreak in northeastern Kenya in 2006–2007 killed more than 420 000 sheep and goats and projected milk losses were estimated to be more than 2.5 million litres due to abortions in cattle and camels.

Because the impact of climate shocks on income and food can be significant, it is crucial that those affected are able to cope with their losses and adapt their livelihoods to deal with changing climate variability and extremes. Identifying the effects of climate shocks on livelihoods and coping and adaptation strategies is key to addressing the impact on food security and nutrition, as will be seen later in a subsequent section.

Impacts on food utilization and food safety

Climate variability and extremes have repercussions for food utilization as they jeopardize the nutritional quality of food produced and consumed, as well as food safety. Although the impacts on food utilization are relatively under-researched – compared with those on availability and access – a number of studies suggest that climate variability negatively affects the nutrient quality and safety of food. In many countries, food variety and diet diversity fluctuate across seasons. Increased inter-seasonal climate variability thus magnifies nutrient intake fluctuations, exacerbating the negative effects on nutrition.

Reduced quality and diversity of diets due to income reductions and high food prices

Households engage in coping strategies in response to food and income reductions and increased prices following climate shocks. This may compromise the quality of the food they consume and the diversity of their diets. Coping strategies that compromise dietary diversity and quality include eating fewer meals per day and less at each meal, skipping meals and eating less nutrient-dense foods and/or more calorie-dense foods high in fat, sugars and salt.

This link between climate shocks, the adoption of coping strategies and the resulting negative impacts on dietary diversity and quality of food are well documented.[146] For example, in Bangladesh climate shocks that affect rice production often lead to higher rice prices, which are strongly associated with greater prevalence of child underweight and poorer dietary diversity.[147] Similar findings are reported for Indonesia in a study conducted at the household level.[148] In both studies, high rice prices negatively affect nutrition mainly through the reduction in the level of expenditures on non-grain food items.

People living in Rufiji, on the coast of the United Republic of Tanzania, have been affected by both prolonged dry seasons and floods. Consequently their eating habits have changed due to a lack of sufficient legumes and fish in the past years, as well as poor harvests of crops due to climate variability and rising food prices. During prolonged dry seasons, this means instead of eating three meals per day, people in Rufiji eat two or even one. New diets include stiff porridge and cooked unripe mangoes.[149]

In many countries, there are seasonal variations in childhood acute malnutrition, where the prevalence increases two- or threefold in the months immediately preceding the harvest. This period often coincides with the rainy season, when food shortages and a lack of dietary diversity combine with a higher incidence of infection.[150] Increased seasonal climate variability often worsens these seasonal variations in acute malnutrition in children.

Food variety, dietary diversity, and food/nutrient intake fluctuate across seasons.[151] Delayed on-set of the growing season or variability in the distribution of seasonal rainfall can worsen fluctuations in food and nutritent intake across seasons. In Malawi, during the lean season – between planting and harvesting – there are substantive decreases in per capita dietary energy consumption and other nutrient acquisition as compared to the post-harvest season. At the individual level, dietary diversity decreased by 26 percent and 30 percent respectively, between the planting and harvesting seasons.[152]

A study carried out in a mountainous area of northern Viet Nam among children aged 24–59 months showed significant seasonal fluctuations in total energy intake: highest in autumn, lower in spring and winter, and lowest in summer. In addition, the intake of carbohydrates, protein and lipids in autumn was higher than in other seasons. Winters are cold and dry (November–February) and summers are rainy and hot (May–August). Spring (March–April) and autumn (September–October) are the two short transitional seasons. Focus group interviews revealed that cool weather in autumn made children eat more than in other seasons.[153]

In some Pacific Small Island Developing States (SIDS), the recurrence of climate shocks that impact on national/local food production, coupled with insufficient recovery time, undermines food security and nutrition in the longer term. This is seen in reduced agricultural and fisheries productivity, increased reliance on short-term humanitarian food assistance, gradual erosion of traditional food systems and intensified permanent shifts away from diversified, healthy traditional diets to greater exposure to imported processed foods often high in salt, sugar and fat. Associated dietary changes heighten the risk of overweight, obesity and diet-related non-communicable diseases (NCDs), as explained next.[154]

Reduced quality and safety of food

More erratic rainfall and higher temperatures along with other extreme events affect the quality and safety of the food in the post-harvest value chain. In relation to safety, higher intensity rainfall is likely to create conditions that lead to mould growth and the subsequent contamination of crops that are still ripening in the field with mycotoxin contamination, toxins that are naturally produced by certain moulds. This is particularly exacerbated in cases where drying efficiencies are lost and where crops are damaged by insects – both highly favourable conditions to a sharply and at times dramatically increased mycotoxin contamination of important staple crops,[155] which eventually renders crops unfit for use as food or as feed. For example, for certain toxins produced by mould (aflatoxins causing liver cancer in all consumers and stunting in children), a combination of drought stress in the pre-harvest period and higher intensity rainfall during harvest and post-harvest stages is ideal for food contamination. Higher temperatures also favour pest and fungi development during storage that can accelerate processes that lead to nutrient deterioration.

Many bacteria, viruses, and parasitic protozoa are strongly climate-dependent and sensitive to climate extremes. Changing climate conditions and extremes such as temperature and humidity alter their survival and transmission patterns and can lead to increased bacterial, viral and pathogenic contamination of water and food. Even increased contamination of water used for irrigation can affect the safety of crops and animals that consume them, as well as their resulting food output.

Unsafe water and food create a vicious cycle of diarrhoea and malnutrition, threatening the nutritional status of the most vulnerable. Where food supplies are insecure, people tend to shift to less healthy diets and consume more "unsafe foods", in which chemical, microbiological and other hazards pose health risks and further aggravate an already marginal nutritional status.[156]

Some food-borne pathogens have survival or multiplication rates sensitive to climate variability and extremes. For example, the multiplication of *Salmonella spp.*, a major contributor to food-borne disease and estimated to be responsible for over 50 000 deaths in 2010,[157] markedly depends on temperature. A recent study indicates that cases of salmonellosis increased by 5.5 percent for each 1 °C increase in mean monthly temperature in Kazakhstan.[158]

Rising sea surface temperature can cause a higher mobilization of heavy metals and is leading to changing patterns and new geographic areas that are affected by harmful algal bloom. The toxins produced by the algae that multiply explosively during an algal bloom often get enriched in the food chain and – though not a threat to fish and seafood themselves – can ultimately cause seafood in the affected areas to be unsafe for human consumption. At a local level, this has direct consequences on coastal communities for which fish can often be the only source of protein. Globally, with seafood being the most internationally traded food commodity, consumers are affected everywhere. While algal bloom has been endemic in certain tropical areas, climate changes cause this to occur more and more frequently in areas that have not previously been affected, where the local population is unprepared to manage such a new threat to their health. Where a concentration builds of heavy metals, they too will accumulate in the food chain and ultimate harm consumers.

In terms of quality, climate extremes can affect the quality of diets through disruption of transport infrastructure, resulting in spoilage and/or reduced access to fresh fruit and vegetables, meat and dairy products. Increasing temperatures and changes in precipitation have already resulted in farmers around the world introducing various climate change adaptation strategies such as crop diversification, mixed crop-livestock farming systems, changing planting and harvesting dates, and using drought-resistant varieties and high-yield water-sensitive crops. While such strategies help maintain food production, the introduction of new crops and cultivation methods also increases the risk of introducing food-borne diseases that people and health systems are not familiar with.[159]

Impacts on health and nutrition

Climate-driven human health impacts are critical to food security and nutrition. As seen in Part 1, disease interferes with the body's ability to absorb nutrients, which can negatively affect the nutritional status of adults and children. Furthermore, recurrent infections and disease are serious contributing factors to both wasting and stunting in children. Disease is also a significant risk factor for impaired maternal nutrition, affecting not only the nutritional status of the mother but also the nutritional status and health of the unborn child. These climate-related negative impacts can undermine a person's ability to work as well as reduce their productivity, which can seriously threaten access to food and income, quality of diet and ultimately food security and nutrition.

Climate variability and extremes can affect human health directly, through changes in temperature and precipitation and natural hazards such as heatwaves, floods, cyclones, droughts; as well as indirectly, through the effect of climate on ecological-mediated risks (e.g. vector-borne and other infectious diseases, crop failures), food safety risks (mycotoxins, heavy metals, harmful algal blooms, etc.) and social responses to climate shocks (e.g. displacement of populations following prolonged drought) (see Figure 35).[160]

Increased health risks and disease
Exposure to more frequent and intense heatwaves is increasing, and the health impacts range from

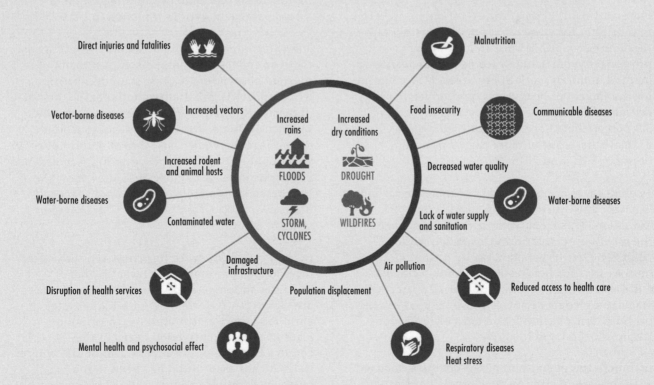

FIGURE 35
HEALTH CONSEQUENCES OF EXTREME CLIMATE-RELATED EVENTS

SOURCE: WHO. 2016. El Niño threatens at least 60 million people in high-risk developing countries. In: *WHO* [online]. Geneva, Switzerland. www.who.int/hac/crises/el-nino/22january2016/en

direct heat stress and heatstroke to worsening of pre-existing conditions such as heart failure, along with a higher incidence of acute kidney injury from dehydration in vulnerable populations. Elderly people, children younger than 12 months, and people with chronic cardiovascular and renal disease are particularly sensitive to these changes.[161] An estimated 125 million additional vulnerable adults were exposed to heatwaves between 2000 and 2016, with a record 175 million people exposed to heatwaves in 2015.[162]

Heatwaves can increase morbidity and mortality associated with heat stress and people with obesity and diet-related NCDs (diabetes, hypertension and cardiovascular disease) are at higher risk. During the 2003 European heatwave, mortality rates among people with cardiovascular disease were 30 percent higher and there were 30 percent more in-patient admissions than comparable periods without heatwaves. Fatal heatstroke occurs 3.5 times more frequently in overweight or obese adults than normal-weight adults. [163]

High and rising temperatures not only pose a risk to mortality for vulnerable populations but also threaten occupational health and labour productivity, particularly for people undertaking manual, outdoor work in hot areas.[164] Accounting for the impact of heat stress on productivity, it is estimated that labour capacity diminished by 5.3 percent between 2000 and 2016, with a dramatic decrease of more than 2 percent between 2015 and 2016.[165]

Though there are some peaks of increased labour capacity, the overwhelming trend is one of reduction. This trend is most notable in some of the most vulnerable countries in the world (Figure 36). Loss of labour capacity has important implications for the livelihoods of individuals, families and communities, potentially affecting wage and income opportunities for those relying on subsistence farming and agricultural wage labour for food and income.

Although the global number of deaths associated with infectious diseases has largely decreased overall since 1990,[166] changing climate variability poses a challenge given the significant association between increasing temperatures,

rainfall and humidity and the rise in disease in many countries. Water-borne diseases and vector-borne and zoonotic diseases have both shown to be sensitive to climate variability and extremes and are significantly associated with the nutritional status of children.[167]

Extreme water-related events make water-borne disease outbreaks more probable. Water-borne disease outbreaks are most commonly a result of excessive precipitation (55 percent of outbreaks) and floods (53 percent) as well as the subsequent contamination of the drinking water supply.[168] Multiple epidemiological studies also have linked El Niño events with increased incidence of disease in human populations. For example, in both rural and urban locations in Bangladesh, cases of cholera and shigellosis rise following greater monsoon flooding and higher sea temperatures as a result of El Niño. Single-study associations between climate variability and extremes and higher disease incidence have been reported for other diseases, including hepatitis A in Australia; dysentery in eastern China; and Bartonellosis, dermatological infection, and Vibrio parahaemolyticus infection in Peru.[169]

Diarrhoeal diseases are particularly worrying as they can reduce food intake and diminish nutrient absorption, leading to undernutrition, while underlying malnutrition increases the risk of diarrhoeal disease.[170] Greater frequency and severity of floods and droughts can exacerbate the occurrence of diseases, due to deterioration in water quality, water scarcity, and higher burdens of malnutrition. A number of studies show the link between climate variability and seasonal diarrhoea, particularly among children under five years of age.

In the northwestern Amhara region of Ethiopia, for example, a recent study revealed that increases in temperature and rainfall in the area are significantly correlated with higher rates of childhood diarrhoea morbidity, the second leading cause of childhood death in the country.[171] Another example is Cambodia, where a significant association has been found between flooding and increased diarrhoea cases in children. Given the two-way interaction between nutrition and diarrhoeal disease and the fact that malnutrition is already a public

FIGURE 36
LABOUR CAPACITY LOSS DUE TO EXTREME HEAT EXPOSURE (CHANGE IN 2006–2016 RELATIVE TO 1986–2008)

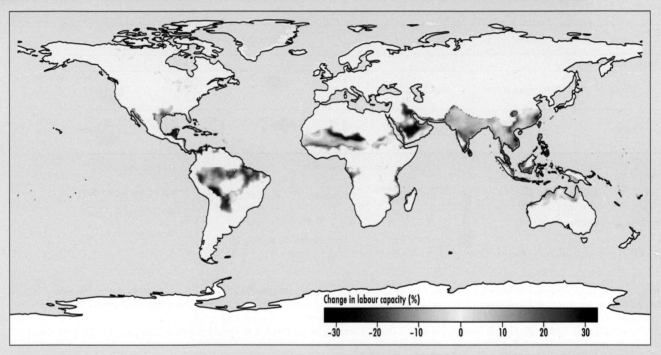

NOTES: Estimated using wet bulb globe temperatures as the change in outdoor labour productivity as a percentage relative to the reference period (1986–2008). The time series of global mean temperatures is used, calculated from the gridded data and weighted by area (to avoid bias from measurements near the poles) and by exposure (to show the number of people exposed).
SOURCE: N. Watts, M. Amann, S. Ayeb-Karlsson, K. Belesova, T. Bouley, M. Boykoff, P. Byass *et al.* 2018. The Lancet Countdown on health and climate change: from 25 years of inaction to a global transformation for public health. *The Lancet*, 391(10120): 581–630.

health threat in Cambodia,[172] ever greater climate variability and extremes are significant threats to the health and nutrition of the country's population.

Vector-borne disease (VBDs) – which generally refer to infections transmitted via the bite of blood-sucking arthropods, such as mosquitoes – are sensitive to variations in rainfall, humidity and temperature.[173] These are some of the best-studied diseases associated with climate variability and extremes due to their widespread occurrence and sensitivity to climatic factors.[174] Malaria and dengue are not only the most sensitive VBDs to climatic drivers, but they also have the highest reported impact in terms of health, affecting more than 270 million people per year combined (Figure 37).

Malaria mainly occurs in Africa and South-eastern Asia and is highly sensitive to increases in temperature, rainfall and humidity. There is evidence that El Niño is associated with a greater risk of certain diseases – not only cholera but also malaria – in specific geographical areas.[175] Malaria by far affects the largest number of people, estimated at 220 million cases per year. Although controversial, recent research shows a strong and significant relationship between malaria and malnutrition, especially for children in high transmission areas.[176] The disease can exacerbate iron deficiency anaemia and contribute to maternal anaemia, with substantial risks for pregnant women, foetuses and newborn babies.[177]

Dengue is the most rapidly spreading VBD, showing a thirtyfold increase in global incidence

FIGURE 37
MOSQUITO-BORNE DISEASE INCIDENCE AND SENSITIVITY TO CLIMATE VARIABILITY AND EXTREMES

Disease	Area	Cases per year	Climate sensitivity and confidence in climate effect
Mosquito-borne diseases			
Malaria	Mainly Africa, SE Asia	About 220 million	Temperature > +- 1; Precipitation > +- 1; Precipitation < -; Humidity > +-
Dengue	100 countries, esp. Asia Pacific	About 50 million	Temperature > +; Precipitation > +; Temperature < -; Precipitation < -; Humidity > +-

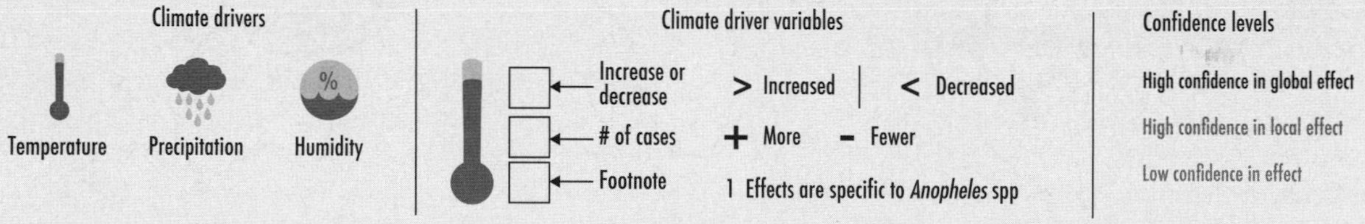

NOTES: Shows the association between different climatic drivers and the global prevalence and geographic distribution of selected vector-borne diseases observed over the period 2008–2012. Among the vector-borne diseases shown, only dengue fever was associated with climate variables at both the global and local level (high confidence), while malaria and haemorrhagic fever with renal syndrome showed a positive association at the local level (high confidence).
SOURCE: Adapted from K.R. Smith, A. Woodward, D. Campbell-Lendrum, D.D. Chadee, Y. Honda, Q. Liu, J.M. Olwoch, B. Revich and R. Sauerborn. 2014. Human health: impacts, adaptation, and co-benefits. In IPCC. 2014. *Climate Change 2014: Impacts, adaptation, and vulnerability. Contribution of Working Group II to the Fifth Assessment Report of the Intergovernmental Panel on Climate Change*, pp. 709–754 [C.B. Field, V.R. Barros, D.J. Dokken, K.J. Mach, M.D. Mastrandrea, T.E. Bilir, M. Chatterjee, K.L. Ebi, Y.O. Estrada, R.C. Genova, B. Girma, E.S. Kissel, A.N. Levy, S. MacCracken, P.R. Mastrandrea and L.L. White, eds]. Cambridge, UK, and New York, USA, Cambridge University Press.

over the past 50 years.[178] It is also the only disease associated with climate variables at both global and local levels at high confidence.[179] Each year there are about 390 million dengue infections worldwide, of which roughly 50 million present symptoms. Three-quarters of the people exposed to dengue are in the Asia-Pacific region, but many other regions are also affected.

Extreme climate variability-related disasters impact on mental health in both the short and long term, with rises in anxiety, depression, post-traumatic stress disorder, chronic distress and incidence of suicide reported.[180] Repeated floods and droughts can also force population displacement – which, in turn, is associated with heightened risks of a wide range of negative health effects. These can include anything from depression to communicable diseases to negative health outcomes caused by civil conflict.[181]

Impacts on women and child care

Women and young children can be particularly vulnerable to climate variability and extremes, as can the elderly and socially isolated.[182] There is valuable, though limited, evidence reporting health impacts for these groups in different countries.

In Viet Nam, the elderly, widows, disabled people, single mothers, and households headed by women with small children were least resilient to floods and storms and slow-onset events such as recurrent droughts.[183] In Bangladesh, according to estimates, women and children represent up to 90 percent of the victims in cyclone-stricken areas.[184] In the aftermath of the 2004 tsunami in Indonesia, Sri Lanka, India and Thailand, a study found that surviving men outnumbered women by almost three to one.[185]

The health impacts that women and children experience through nutrition are even less studied but the relationship exists. Climate variability can undermine maternal and child care and breastfeeding practices, amplifying food shortages in which women consume less food[186] and suffer from reproductive tract infections and water-borne diseases after floods.[187]

The role of women as primary caregivers and providers of food and fuel for households makes them more vulnerable when flooding and droughts occur. In Central Africa, where up to 90 percent of Lake Chad has disappeared, nomadic indigenous groups are especially at risk.[188] As the lake's shoreline recedes, women have to walk much further to collect water. And with dry seasons now becoming longer in many countries in Africa, women are working even harder to feed and care for their families without support.

There is some evidence that climate shocks can increase the workload of women farmers and raise farming-related health risks.[189] This in turn can limit women's ability to follow recommended breastfeeding and complementary feeding practices and offer nutritious food with recommended frequency and responsiveness to their young children.[190]

Breastfeeding protects infants against food- and water-borne illnesses that can be more common after extreme climate events, and also protects against non-communicable disease (NCDs) in later life. When a woman's ability to exclusively breastfeed her infant for six months is reduced, this poses an increased health risk to infants and young children.[191] There is evidence that the effects of climate shocks on child undernutrition may be exacerbated through diminished child feeding and caregiving practices. Furthermore, these effects will be greater in settings where they are combined with pre-existing vulnerabilities related to poor health and malnutrition.[192] ∎

2.3 WHAT ARE THE IMPACTS OF CLIMATE ON THE VULNERABILITY, RESOURCE AND CONTROL FACTORS THAT SHAPE FOOD SECURITY AND NUTRITION?

KEY MESSAGES

➔ Climate variability and extremes have impacts on livelihoods and livelihood assets – especially of the poor – contributing to greater risk of food insecurity and malnutrition.

➔ Climate shocks and environmental degradation reduce goods and services available to people and local communities, not only limiting their economic opportunities and livelihood options but also modifying their resilience, coping and adaptive capacity.

➔ Prolonged or recurrent climate extremes lead to diminished coping capacity, loss of livelihoods, distress migration and destitution.

➔ Climate-related disasters create and sustain poverty, contributing to increased food insecurity and malnutrition as well as current and future vulnerability to climate extremes.

➔ Extreme climate events have short-, medium- and long-term impacts on food security and nutrition.

Climate variability and climate extremes can affect the viability of livelihoods and result in adjustments to livelihood strategies. Repeated climate shocks can undermine households' ability to maintain their livelihood asset base or to reinvest in agriculture, leading some to chronic food insecurity, malnutrition, poor health, and lack of economic productivity. There is evidence that the livelihoods of the poor are particularly affected.[193]

Against this backdrop, a focus on peoples' assets or different types of capital is central not only to understanding the impacts of climate shocks on livelihoods and coping and adaptation strategies, but also to identifying key factors to be considered for policy design and the implementation of programmes aimed at improving food security and nutrition. A focus on assets or capital also helps to establish what resources are available and accessible in order to aid in adaptation.

Impacts on livelihood assets

The analysis on the impact of climate variability and extremes on household and individual assets or different types of capital focuses on five types (natural, physical, human, financial and social), which are defined according to the Sustainable Livelihoods Framework (see Annex 4 Glossary). Understanding how these types of capital are affected in the event of climate shocks sheds further light on expected changes in exposure and vulnerability to climate variability and extremes.

Impacts on natural capital

Climate shocks contribute to environmental degradation. It is well known that climate-related disasters are a significant factor in ecosystem degradation and loss, including increased soil erosion, declining rangeland quality, salinization of soils, deforestation, reduction of quantity and quality of ecosystem services, and biodiversity loss.[194] Consequently, economic opportunities and livelihood options of households who are heavily dependent on natural resources to meet their food security and nutrition needs are also affected by climate shocks.[195]

Higher temperatures and humidity are raising the risk of fungal growth and thus the contamination of stored cereals and pulses with mycotoxins (fungal metabolites). Climate variability and more frequent climate extremes (e.g. dry spells, intense short-lived widespread rainfall, and cyclones), in addition to causing severe disruption in their own right, can lead to more frequent and intense plant pest and disease outbreaks. This was the case during the desert locust outbreaks both in north-western Africa and in Yemen in late 2015 and early 2016.[196]

Unfortunately, the impact of climate extremes on natural resources and the environment remains a largely neglected area in terms of direct and indirect economic losses. Although there is a wide range of studies that look at the climate effects on soils, most of these overlook links to agriculture, food security and nutrition, partly because of the lack of reliable data. This gap is being addressed with new developments in global soils data[197] as well as a comprehensive review of the impacts on ecosystem services for food production.

Impacts on physical capital

The physical damage caused by climate-related disasters has direct impacts on agriculture and the food value chain. These can come in the form of disruption to the flow of agricultural inputs such as seeds and fertilizers or in challenges to processing and distribution, markets, retailers, and final consumption.

Floods and other climate-related disasters can potentially undermine fishing and damage aquaculture infrastructure and facilities such as fish farms, fish ponds, oyster banks, fish feed storage, fish reproduction facilities, boats and gear. This will result in major losses in fish and aquaculture production and livelihoods.

In Pakistan, heavy monsoons caused floods in 2010 that destroyed property, assets and infrastructure, affecting millions of people.[198] Small to medium-sized agribusinesses were hurt in cotton ginning, rice processing, flour and sugar milling, silk and horticulture. There was also damage to agriculture infrastructure, including machinery, warehouses, irrigation systems, animal health clinics, agriculture and livestock research and extension offices, and government buildings and facilities.[199] Cyclone Nargis, which struck Myanmar in 2008, caused havoc to forestry, fishery and agriculture. Over half of small rice mills and two-thirds of larger rice mills in the affected areas were damaged, and losses in terms of farm machinery and land affected the wider 2008/09 rice crop.

Such damage and destruction of physical capital undoubtedly affect the quality of diets and food stability. For example, disruption of transport infrastructure due to extreme climate often

FIGURE 38
HEALTH AND EDUCATION FACILITIES DAMAGED BY DISASTER TYPE, 1994–2013

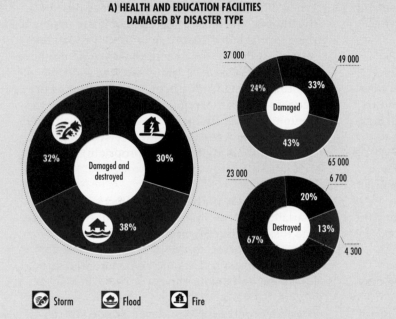

A) HEALTH AND EDUCATION FACILITIES DAMAGED BY DISASTER TYPE

Damaged and destroyed: Storm 32%, Fire 30%, Flood 38%

Damaged: 37 000 (24%), 49 000 (33%), 65 000 (43%)

Destroyed: 23 000 (67%), 6 700 (20%), 4 300 (13%)

Storm | Flood | Fire

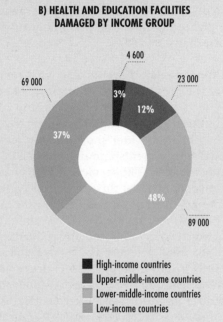

B) HEALTH AND EDUCATION FACILITIES DAMAGED BY INCOME GROUP

4 600 (3%), 23 000 (12%), 89 000 (48%), 69 000 (37%)

- High-income countries
- Upper-middle-income countries
- Lower-middle-income countries
- Low-income countries

NOTE: Percentage of health and education facilities damaged by three types of natural disasters: storms, floods and fires.
SOURCE: Centre for Research on the Epidemiology of Disasters (CRED). 2015. *The human cost of climate-related disasters: a global perspective 2015.* Data are from Emergency Events Database (EM-DAT). 2009. *EM-DAT* [online] Brussels. www.emdat.be

results in spoilage or reduced access to fresh produce, meat and dairy products, thus potentially affecting diet quality and food safety. Fruits and vegetables are more challenging to produce and distribute, as they are not only vulnerable to extreme climate conditions but also to any disruption in the transport/storage/cold chain infrastructure.[200]

Impacts on human capital

When extreme climate events damage other infrastructure not necessarily associated with food value chains, such as health and education facilities, there can also be considerable impacts on human capital, including health and nutrition. In the long term, the loss of education and health infrastructure can be detrimental to the achievement of universal health coverage, economic growth and social development for generations, with negative impacts on food security and nutrition.

Alarmingly, more than 185 000 health and education facilities were either damaged or destroyed worldwide by climate-related disasters between 1994 and 2013. Floods were the leading cause of damage, followed by storms (Figure 38).[201] In the overwhelming majority of cases – 85 percent – this damage occurred in low- and lower-middle-income countries. These countries already face significant challenges in the provision of universal coverage for adequate health and education services and have limited capacity to rebuild in the aftermath of climate-related disasters.

The damage to health facilities disrupts the provision of health services that are especially critical during and after disasters. Many studies have shown that the health and nutritional status of children in particular is especially vulnerable to disaster, both during the emergency phase and – due to malnutrition and undernutrition – in the aftermath.[202]

Climate-related events also have consequences for the provision and operation of health services. Indeed, a food security assessment conducted in southern Africa found that in early 2016 (during the El Niño), water shortages limited access to health treatments and disrupted HIV and tuberculosis (TB) services.[203] This has serious consequences, as this particular region accounts for one-third of all people living with HIV (PLHIV) worldwide. PLHIV are highly dependent on nutritious food, and any reductions in food intake may decrease the effectiveness of antiretroviral therapy (ART) drugs and also treatment adherence. Further, poor nutrition may reduce immunity and increase risk for HIV infected children lacking ART and can also result in malnutrition and infections for TB patients.

The damage or destruction of any type of capital, whether it be natural resources, physical capital or human capital, is important in its own right. Nonetheless, climate shocks can be such that damage or destruction can befall various types of capital at the same time. When this is the case – e.g. for Small Island Developing States (SIDS) – there can be serious long-term implications for increasing malnutrition in all its forms and for non-communicable disease (NCD) (Box 10).

Impacts on financial capital

Financial assets play a key role in enhancing the resilience of vulnerable groups. A reduction in financial capital weakens adaptive capacities of households and increases their vulnerability. This is quite relevant for climate resilience when considering the negative impacts of climate variability and extremes on agricultural production, yields and income (identified earlier).

Low crop yields stand out as a potential stressor on people's financial capital.[204] Seasonal crop failures also lead to high food prices and push households to spend a larger proportion of their income on food. This has the potential to affect the quality of household diets and magnify the risk of malnutrition, while also leading to a loss in household financial capital.

When climate variability and extremes disrupt livelihoods, the most affected people are unable to raise formal bank loans due to lack of collateral (often lost during the event) and often do not have insurance.[205] Diseases and other stressors on health resulting from climate shocks often restrict people's ability to work and therefore impede the accumulation of financial capital. Poor health and difficulties in accessing health services limit households' ability to seek appropriate health care, also affecting ability to work.

As noted when analysing the effects on food availability, more than 80 percent of the damage and losses caused by droughts affect the agriculture sector, not only in crop production but also livestock. This includes potential animal losses due to climate shocks.

For many rural people in developing countries livestock can be part of a financial strategy or a coping mechanism, representing an important asset to generate financial capital. In rural areas of many low- and middle-income countries, financial services such as credit, banking and insurance are virtually non-existent. In these areas, livestock plays an important role as a means of saving and capital investment, often providing a substantially higher return than alternative investments.[206] Investments in livestock are also used to hedge against rapid inflation, as well as against unexpected climate-related disasters such as droughts and floods.[207]

In Somalia, for example, livestock acts as a "bank on four legs" used to access cash, and herds serve as valuable trade items exchanged for food and other essentials. However, three years of drought have taken a heavy toll on livestock. Losses of goats, camels, sheep and cattle in 2017 alone ranged from 20 to 40 percent – reaching 60 percent in the hardest-hit drought locations. These large-scale livestock deaths undermine the viability of livelihoods and push families over the edge, leading to high levels of food insecurity and malnutrition and forced economic displacement as families search for relief.[208]

BOX 10
SMALL ISLAND DEVELOPING STATES: DESTRUCTION OF NATURAL, PHYSICAL AND HUMAN CAPITAL AND LONG-TERM IMPLICATIONS FOR NON-COMMUNICABLE DISEASE AND MALNUTRITION

Geography and socio-economic characteristics in the Pacific render Small Island Developing States (SIDS) particularly vulnerable to tropical cyclones, droughts and floods. Worldwide, 5 of the 15 countries considered the most vulnerable to natural hazards are Pacific SIDS, with Vanuatu ranked as the most vulnerable globally.

In 2015 a category 5 tropical storm, Cyclone Pam, caused widespread devastation in Vanuatu, affecting approximately 200 000 people (73 percent of the population) and causing an estimated USD 590 million in damages (65 percent of GDP).[1] Reefs were damaged and fishing infrastructure destroyed. Additionally, 70 percent of food crops were destroyed, putting more pressure on already-declining fisheries for local consumption.[2] Cyclone Pam was followed only months later by a severe El Niño-induced drought that exacerbated the impacts of the cyclone, impeded recovery and resulted in further crop losses and water scarcity.[3]

These effects reinforce the already ongoing dietary transition away from a healthy traditional local diet to a greater dependency on imported foods and beverages, often high in fat, sugar and salt, leading to an increase in overweight, obesity and diet-related

non-communicable disease (NCDs). The enabling processes that underpin nutrition status and development outcomes (e.g. political commitment, policy environment for action and implementation) are also undermined by climate shocks in SIDS, where already-limited government capacity is further stretched, long-term vision is impaired and focus is directed to the immediate needs following a shock.

While climate shocks can rapidly increase acute malnutrition (wasting), micronutrient deficiencies and prevalence of infectious diseases in the short term, the longer-term impacts on nutrition and health status should not be overlooked. Expected to increase in frequency and intensity in the Pacific, climate shocks affect the immediate, underlying and enabling processes that determine nutrition and can thus reinforce all forms of malnutrition, including overweight and obesity and diet-related NCD. A teacher on Emae Island described the impact of Cyclone Pam on food availability, water security and education as follows:

"After cyclone Pam, the water that belongs to you and me was not very good. I had to stop class sometimes, half days, and then we'd eat all together, sometimes we tell the children not to come to school tomorrow because we don't have enough food."[4]

SOURCES:
[1] F. Thomalla and M. Boyland. 2017. *Enhancing resilience to extreme climate events: lessons from the 2015-16 El Nino event in the Asia Pacific.* Stockholm, Stockholm Environment Institute.
[2] Food Security and Agriculture Cluster. 2015. *Vanuatu Food Security & Agriculture Cluster CYCLONE PAM Medium and Long Term Recovery and Rehabilitation Strategy 2015–2017.*
[3] United Nations Office for the Coordination of Humanitarian Affairs. El Nino in Vanuatu 2015 [available at https://reliefweb.int/report/vanuatu/el-ni-o-vanuatu].
[4] G. Jackson, K. McNamara and B. Witt. 2017. A Framework for Disaster Vulnerability in a Small Island in the Southwest Pacific: A Case Study of Emae Island, Vanuatu. *International Journal of Disaster Risk Science,* 8(4): 358–373.

In Zimbabwe, relatively wealthier households sell livestock to smooth consumption in the face of drought-induced agricultural income losses, whereas poorer households cope with income losses by smoothing assets through decreased consumption.[209] The latter raises issues of food security and nutrition threats for the poor: the erosion of assets (e.g. livestock) makes them more exposed to future risks.

Fisheries assets used to generate financial capital are also highly vulnerable, particularly in the face of storms and hurricanes. Hurricane Gilbert in 1998 was particularly damaging, with Jamaican fishers losing 90 percent of their traps. This meant a loss in revenue, and costly repairs and delays in resuming fishing activities.[210] In Peru, at the time of the 1997–1998 El Niño, a percentage of the catch

BOX 11
SEVERE DROUGHTS CAN CONTRIBUTE TO INCREASED SOCIAL INSTABILITY AND TRIGGER CONFLICTS

Drought can threaten local food security and nutrition and aggravate humanitarian conditions, which can trigger large-scale human displacement and create a breeding ground for conflict. Some studies indicate that, as drought intensifies and lingers, the likelihood of conflict rises significantly.[1]

In agriculture-dependent communities in low-income contexts, droughts have been found to increase the likelihood of violence and prolonged conflict at the local level, which can eventually pose a threat to societal stability and peace.[2]

Some examples include:

▶ persistent drought in Morocco during the early 1980s, which resulted in food riots and contributed to a macroeconomic collapse;[3]

▶ drought in the Syrian Arab Republic between 2006 and 2010, which affected 1.3 million people, accelerating rural migration to cities and compounding other stresses and sources of tension;[4]

▶ a drought in Somalia that fuelled conflict through livestock price changes, establishing livestock markets as the primary channel of impact;[5]

▶ cattle raiding as a normal means of restocking during drought in the Great Horn of Africa (GHA), which then leads to conflict;[6] and

▶ a region-wide drought in northern Mali in 2012, which wiped out thousands of livestock and devastated the livelihoods of pastoralists, in turn swelling the ranks of armed rebel factions and forcing others to steal and loot for survival.[7]

SOURCES:

[1] J.F. Maystadt and O. Ecker. 2014. Extreme weather and civil war: does drought fuel conflict in Somalia through livestock price shocks? *American Journal of Agricultural Economics*, 96(4): 1157–1182.

[2] FAO, IFAD, UNICEF, WFP and WHO. 2017. *The State of Food Security and Nutrition in the World 2017. Building resilience for peace and food security.* Rome, FAO.

[3] H. El-Said and J. Harrigan. 2014. Economic Reform, Social Welfare, and Instability: Jordan, Egypt, Morocco, and Tunisia, 1983–2004. *The Middle East Journal*, 68(1): 99–121.

[4] WFP and ODI. 2015. Food in an uncertain future: The impacts of climate change on food security and nutrition in the Middle East and North Africa. Cairo, WFP and London, ODI.

[5] Maystadt and Ecker, 2014 (see source 1).

[6] IGAD Climate Prediction & Applications Centre (ICPAC) and WFP. 2017. *Greater Horn of Africa Climate Risk and Food Security Atlas.* Nairobi.

[7] C. Breisinger, O. Ecker and J.F. Trinh Tan. 2015. Conflict and food insecurity: How do we break the links? In IFPRI, eds. *Global Food Policy Report 2014–2015*, pp. 51–59. Washington, DC.

value was put into a recently privatized social security and health organization for industrial fishers. However, as a result of decreasing catches, the agency's coffers quickly ran dry.[211] This left fishers without a safety net and access to financial resources to cope with the difficult economic situation.

Impacts on social capital

There is mounting evidence that climate-related disasters also diminish social capital, thereby reducing people's adaptive capacities. Social capital builds upon institutions embedded in social structures and relations that lead to trust, improved information exchange, lower transaction costs and the likelihood of collective action.[212]

Last year, this report presented evidence that climate-related events, especially drought (see Box 11), can become a trigger for social instability and violence, as they tend to jeopardize food security, which in turn has been found to increase the risk of conflict.[213] This is particularly the case where deep divisions exist between population groups, in contexts of pervasive inequality and fragile institutions and where coping mechanisms are lacking.

Some studies find that deviations from moderate temperatures and precipitation patterns systematically raise conflict risk. Temperature has the largest average impact, with each 1 °C rise in temperature increasing conflicts between individuals by 2.4 percent and conflicts between groups of individuals – e.g. organized violence, civil conflicts and riots – by 11.3 percent.[214]

Climate shocks also contribute to environmental degradation and loss; this too can trigger increased competition and become a flash point for unrest, insecurity and conflict. In the Greater Horn of Africa, for example, water, forests and rangelands are becoming more degraded due to a combination of overuse, recurrent droughts and increased temperatures.[215] As a result, competition over scarce pasture and water among pastoral communities often becomes fierce, particularly during drought years when pastoralists are forced to use non-traditional migration routes. During the recent strong 2015–2016 El Niño-related drought, pastoralists were forced to move their herds far beyond their normal grazing areas to nature reserves and farmland in Kenya, where they clashed with local populations there.

Coping strategies adopted by households

The analysis until now shows that climate shocks can undermine a household's ability to maintain its livelihood asset base or to reinvest in agriculture. This interaction between climate events and vulnerabilities determines the basic outcome with regard to food security and nutrition.

The impacts on food security and nutrition can be significant, and people may react in a variety of ways. The analysis that follows focuses on how people cope with a shortfall in food or income following a climate shock (*ex post*), as well as how they adapt their livelihood strategies (*ex ante*) in the context of climate variability.

Resilience is an important factor in coping with the impacts of climate variability and extremes and ensuring that they do not have long-lasting

consequences for food security and nutrition, as previous editions of this report show.[216] There are three capacity types that determine the ways and extent to which individuals, households and communities are able to cope with and adapt to climate shocks and their impact:

▶ adaptive capacity (coping strategies, risk management, and savings);
▶ absorptive capacity (use of assets, attitudes/motivation, livelihood diversification and human capital); and
▶ transformative capacity (governance mechanisms, policies/regulations, infrastructure, community networks and formal safety nets).

Ex post coping strategies

The adoption of coping strategies depends on the nature of the climate shock and the degree of impact on household access to food and income. Strategies can take the form of consumption coping strategies (e.g. skipping meals, switching to cheaper foods, borrowing food, begging) or livelihood coping strategies (selling assets, sending household members to work off-farm, etc.).

Households typically engage first in reversible coping strategies with short-term effects, such as making modest dietary adjustments and skipping meals. However, as coping options are exhausted and food security worsens, households are more likely to employ more extreme and damaging strategies that are less reversible, such as selling productive assets. In the most severe form, a climate shock can lead to the collapse of coping mechanisms entirely and the loss of livelihoods, prompting migration and destitution, and, in the most severe form, starvation and death. In other cases, adopting negative coping strategies results in increased acute malnutrition and stunting among preschool children as a consequence of reduced food access, limited adequate child care and increased exposure to contaminants.[217]

There are many examples where adopting *ex post* coping strategies are detrimental to food security and nutrition (see Box 12). In some contexts, climate shocks can force vulnerable groups to adopt other types of negative coping strategies, such as illegal activities, which are detrimental »

BOX 12
COMMONLY USED *EX POST* COPING STRATEGIES THAT ARE DETRIMENTAL TO FOOD SECURITY AND NUTRITION: SELECTED COUNTRY EXAMPLES

▶ The **Karamoja region in Uganda** is characterized by chronic food insecurity due to high levels of poverty, low development and unfavourable climate conditions. The most frequent coping strategies adopted by households after a climate shock include begging, borrowing, sale of local brew and charcoal/fuelwood production. The selling of assets – particularly livestock – is a commonly employed coping strategy among households in response to droughts/prolonged dry spells.[1]

▶ In **Kyrgyzstan**, reducing consumption quality is the coping strategy most frequently employed by households to mitigate the impact of food security-related climate shocks (see figure below). This indicates that the quality of

consumption is highly sensitive to external shocks, such as climate shocks. This could result in micronutrient deficiencies, thereby jeopardizing the nutrition status of vulnerable household members. Moreover, evidence indicates that rural households are more at risk of food insecurity.[2]

▶ In **Timor-Leste**, drought-affected households have been adopting negative coping strategies such as limiting portion sizes, reducing the number of meals a day, using food stocks necessary for the lean season, and selling household assets. Considering the already low resilience levels of many households in areas worst hit by the El Niño drought, these negative coping mechanisms have further exacerbated fragile livelihoods.[3]

TYPES OF COPING STRATEGIES EMPLOYED BY HOUSEHOLDS WHILE FACING FOOD SHORTAGES IN KYRGYZSTAN

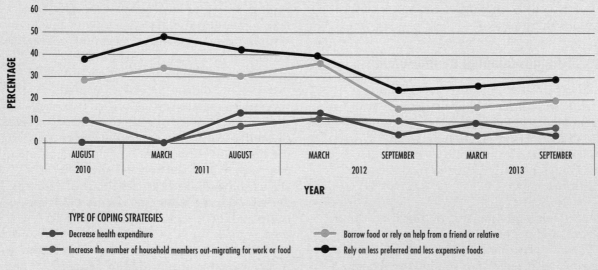

TYPE OF COPING STRATEGIES
● Decrease health expenditure
● Increase the number of household members out-migrating for work or food
● Borrow food or rely on help from a friend or relative
● Rely on less preferred and less expensive foods

SOURCE: WFP. 2014. *Kyrgyz Republic – An overview of climate trends and the impact on food security.* Bishkek.

SOURCES:
[1] IGAD Climate Prediction & Applications Centre (ICPAC) and WFP. 2017. *Greater Horn of Africa Climate Risk and Food Security Atlas.* Nairobi.
[2] WFP. 2014. *Kyrgyz Republic – An overview of climate trends and the impact on food security.* Bishkek.
[3] CARE, Oxfam, PLAN International and World Vision. 2016. *Humanitarian partnership agreement (HPA) agency assessment on El Nino impacts in Timor-Leste.*

» to the well-being of society, as observed for example in the north-eastern zone of Nigeria,[218] and in Guatemala, El Salvador, and Honduras in Central America.[219]

Ex ante adaptation strategies

Not all households take action or even precautionary responses in the face of climate variability and extremes. They could perceive that the stressor is not critical (i.e. feel the opportunity cost of acting is high) or simply lack the means to adapt.[220]

Evidence suggests that the opportunity cost associated with climatic uncertainty is substantial – perhaps greater than the direct, *ex post* cost of shocks.[221] Climate risks affect the behaviour of people, who may reduce their investments and assets because of the possibility of further losses. As a consequence, individuals hit by shocks may opt for lower-risk but lower-return activities.[222]

One of the primary sources of agricultural income risk is production uncertainty caused by climate-related events. Households deplete their productive assets to subsist during transitory shocks,[223] opting for low-risk, low-return investments to mitigate risk over time.[224] Farmers' precautionary strategies include selection of less risky but less profitable crops and cultivars, shifting household labour to less profitable off-farm activities, and avoiding investment in production assets and improved technology.[225]

Problems of access to social and financial services are among the factors that limit households in adopting more long-term sustainable strategies to face climate variability. A lack of formal institutions to reduce household vulnerability to agricultural income risk restricts many countries' ability to cope and adapt both in the short and long term.

Examples of barriers to adaptation cited by farmers include lack of access to credit in South Africa and lack of access to land, information and credit in Ethiopia.[226] Many regions in sub-Saharan Africa are heavily constrained by their limited social, political and technical resources, which already affect their ability to

cope with issues of scarcity and poverty. These constraints also hamper their ability to cope with and adapt to changing environmental conditions.[227]

Nevertheless, with farmers already more consciously noticing changes in rainfall and seasonality,[228] some of them are now using a variety of strategies to adjust or adapt to changes in their environment, despite the aforementioned constraints.

Ex ante diversification strategies help farmers to smooth income streams over time.[229] Climate variability and frequent climate shocks increase incentives to adopt climate-smart agricultural practices such as: the use of drought-tolerant crop varieties; soil and water conservation techniques that restore degraded lands and store water in the soil; and agroforestry technologies that restore soil fertility and control soil erosion and desertification.[230] The type of strategies currently being adopted by households and the conditions that facilitate their implementation are discussed next.

In response to changing rainfall patterns and shorter growing seasons, some farmers are shifting to drought-tolerant crops and fast-maturing varieties in order to adapt.[231] As seen above, these shifts are sometimes aided by social capital – such as government programmes and extension, or communication and support among farmers[232] – demonstrating the important role of higher-level structures and processes. Farmers are also changing planting dates (adjustment of cropping calendars) in response to erratic rainfall or false starts to the rainy season and implementing mixed cropping and crop switching to reduce the risk of total crop failure.[233]

Other changes in farming practices due to changes in rainfall patterns include increasing planting distances in response to soil moisture deficits, introducing short-maturing varieties of maize in response to reduced rainfall at the end of the growing season and the construction of stone bunds to curb soil erosion caused by more intense rainfall.[234] Farmers also draw upon their social capital to build their adaptation strategies. They form cooperatives to reduce production and transportation costs, thus enhancing social capital.

In the Philippines, for example, more intense typhoons have important consequences for food security. They have significant negative effects on households that depend on farming livelihoods.[235] And there is also evidence that El Niño has been equally destructive by lowering rainfall in some Philippine regions with severe impacts on incomes, affordability of food, livelihoods, nutrition and dietary diversity.[236]

In response to these climate shocks, households of landless agricultural workers in a number of agricultural activities have employed various coping mechanisms or survival strategies, including the participation in different kinds of work groups as well as cooperation (whether within or between families) as a form of "shared poverty" (i.e. pooling together of labour and other resources to maximize income).[237] Some of these farmers have also engaged in diversified income strategies and sought alternative livelihood sources, such as carpentry, gardening, raising livestock, vending, construction work or domestic help (both at home and abroad).

Of course, farmers in cooperatives alone cannot go very far. In some cases, their success may also depend on help from government extension agencies to gain access to drought-resistant crop varieties and indigenous livestock breeds.[238]

There are also other measures that some farmers are implementing to a lesser extent, such as reforesting along the banks of water bodies (to prevent soil erosion, reduce water temperature or provide windbreaks for crops), using irrigation and investing in water harvesting schemes, as well as soil and water conservation measurements.[239] Tree planting is also reportedly being implemented, mostly by livestock farmers to protect livestock against heat stress.[240]

In Malawi, the Niger and Zambia, climate variability and extremes can act as push factors for crop and income diversification.[241] In north-eastern Ghana, prolonged dry seasons lead farmers to seek more off-farm employment.[242] In South Africa, a short-term adaptation strategy to

dry spells is to shift from cropping to livestock management.[243] While this strategy is effective in reducing reliance on crops that may fail due to lack of rain, farmers are noting a reduction in grazing resources as a result of this shift.

Internal migration, whether seasonal or more lasting, has also been identified as one of the key coping strategies used by households to diversify income in response to climate shocks and as a risk mitigation strategy.[244] Many scholars consider this a traditional strategy that gives individuals the chance to diversify their income, diversify risk for their household and send money back to family members, thereby boosting resilience back home.[245] Migration pools or avoids risks across space and is especially successful when combined with clear information about potential precipitation failures.[246]

When financing the relocation of a household member within a country is more affordable than other alternatives, migration offers poor households a potential risk management strategy. Moreover, households target destinations where income risk is least correlated with risk at home.[247]

In northern Nigeria, households facing greater *ex ante* risk have a greater probability of having at least one migrant.[248] In the United Republic of Tanzania, for an average rural household, a 1 percent reduction in agricultural income induced by climate shocks increases the probability of migration by 13 percent on average within the following year. However, this effect is significant only for households in the middle tier of wealth distribution, suggesting that the choice of migration as an adaptation strategy depends on initial endowment. This is not necessarily the case when income is highly dependent on agriculture.[249]

In conclusion, some farmers are already taking measures to deal with climate variability and extremes. The adoption of *ex post* adjustments following climate extreme episodes depends on the nature of the event and the degree of impact on the household's access to food and income. It also depends to a significant extent on people's access to extension services, information, credit, savings and livelihood options. Without clear

FIGURE 39
THE GLOBAL SCALE OF DISPLACEMENT CAUSED BY DISASTERS, 2008–2014

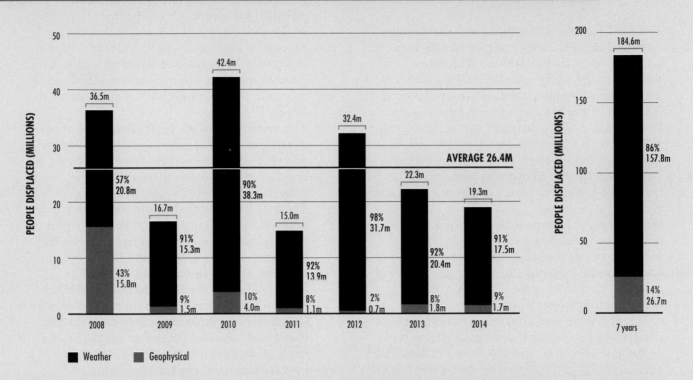

NOTES: Total number and percentage of people displaced between 2008 and 2014 by two broad category types of disaster: weather and geophysical. Following the classification system adopted by the international disaster database (EM-DAT), geophysical events include earthquakes, mass movements and volcanic activity; weather includes meteorological (storms, extreme temperatures), hydrological (floods, landslides, wave action) and climatological events (droughts, wildfires). Differences in total are due to rounding of figures to the nearest decimal point.
SOURCE: Global Estimates (2015), data as of June 2015 from Internal Displacement Monitoring Centre (IDMC).

sustainability criteria – requiring intervention and policy coherence – coping strategies can have detrimental effects. In most cases, households' immediate response to climate variability and extremes can be detrimental to food security and nutrition because the quality of diet consumed is highly sensitive to external shocks, such as increased food prices and climate-related disasters.

When coping and adaptation strategies are no longer an option
In the most severe form, extreme climate events or prolonged/recurrent climate variability can lead to the collapse of coping mechanisms and the loss of livelihoods. This can prompt migration and destitution due to distress when

people have no other viable option to sustain their livelihoods, potentially leading to starvation and death.

In fact, extreme climate shocks can be a significant driver of migration and forced displacement (Figure 39). Disasters brought on by climate-related hazards forced more than 17.5 million people to leave their homes in 2014.[250]

Most displacements induced by rapid-onset events are short-distance and involve temporary movements.[251] However, where there are recurrent climate shocks, patterns of movement can become cyclical, pre-emptive and permanent as a result of perceived future risk. In Bangladesh, approximately 22 percent of rural households affected by tidal-surge floods and 16 percent of those affected by riverbank erosion migrated to urban areas.[252]

The 2011 East Africa Drought and the Somalia Famine 2011–2012 are examples of extreme climate events that, combined with other vulnerability factors – conflict, rising global food prices and other longstanding structural factors – led to the collapse of coping mechanisms and livelihoods, causing destitution and catastrophic levels of food insecurity and malnutrition. These events resulted in a severe food crisis across Djibouti, Ethiopia, Kenya and Somalia, threatening the livelihoods of 9.5 million people.

Many refugees from southern Somalia fled to neighbouring countries such as Kenya and Ethiopia, where crowded, unsanitary conditions and severe malnutrition led to a large number of deaths. Other countries in Eastern Africa, including the Sudan, South Sudan and parts of Uganda, were also affected. A famine was declared in two regions in the southern part of Somalia in July 2012, the first time a famine had been declared in the region by the United Nations in nearly 30 years. Tragically, tens of thousands of people are believed to have died in southern Somalia before famine was even declared.[253]

Exposure and vulnerability of livelihoods and population groups

The analysis presented so far suggests that climate variability and extremes undermine food security and nutrition. The precise impacts depend on people's exposure to climate shocks and their vulnerability to these shocks. Vulnerability here refers to an inability to cope with external changes including avoiding harm when exposed to a hazard. This includes inability to avoid the hazard or shock; anticipate it; take measures to avoid it or limit its impact; cope with it; and recover from it.[254] The evidence shows that low- and middle-income countries are increasingly exposed to climate extremes and their vulnerability to these events is becoming a more important risk factor for food security and nutrition.

Generally, a stress or shock can be amplified or reduced depending on the vulnerabilities at each level of the system. In many cases, climate shocks and risks can be amplified by:

▶ **Environmental, social, economic and political stressors**, which together impinge on livelihoods and reinforce each other in the process, often negatively.[255] Vulnerabilities are in some cases also exacerbated by a lack of education and healthcare facilities, leading to economic impediments with long-term effects.[256]

▶ **The repetition of such stressors and shocks over time**, which erodes households' assets and their capacity to cope. For example, a drought can increase vulnerability to subsequent droughts by: (i) weakening livestock, making them more vulnerable to diseases; or (ii) hampering food production, forcing households to adopt negative coping strategies such as selling or reducing assets.

▶ **Limited ability to cope and adapt** if households lack the right means, so that climate shocks may contribute to even greater vulnerability. Maladaptive actions, or actions that undermine the long-term sustainability of livelihoods, result in downward trajectories, poverty traps and worsening inequality.[257]

▶ **Poverty and persistent inequality**, among the most salient conditions that shape climate-related vulnerability.[258] They reinforce the conditions in which people have few assets to liquidate in times of hardship or crisis.[259] The poor are the first to experience asset erosion, poverty traps and barriers and limits to adaptation.[260] Climate-related disasters also keep people in or move them back into poverty and are one reason that eradicating poverty is so difficult. For example, between 2006 and 2011, 45 percent of poor households in Senegal escaped poverty, but 40 percent of non-poor households became poor, leaving the poverty rate almost unchanged.[261]

▶ **Marginalization**, a critical determinant because vulnerability and adaptation to climate shocks depend on opportunities governed by the complex interplay of social relationships, institutions, organizations and policies.[262] The socially and economically disadvantaged and the marginalized are disproportionately affected by the impacts of climate variability and extreme events.[263]

Therefore, the impact or risk of impact from a climate shock is context-specific – not only will it depend on the nature and intensity of the shock, but also on the fragility of a system or livelihood in relation to this hazard.[264] Moreover, livelihoods are also affected differently by various climate shocks and stressors, depending on the types of livelihoods (if based on crop, livestock, fish, tree, other renewable natural resources, or any combination of these) and their ability to withstand impacts of drought, floods or storms.

For these reasons – nature and intensity of the shock, fragility of a system/livelihood and livelihood type – some livelihoods and population groups are more vulnerable and at greater risk of increased food insecurity and malnutrition. Effects on specific groups include:

▶ **Small family farms and agriculture labourers:** The majority of the world's poor and food-insecure people are rural, either farmers, fisherfolk, or labourers with direct or indirect dependence on agriculture for their income. They are thus directly exposed to any risk that would impact agricultural production. Small family farms are particularly vulnerable. For example, a small family farm that typically relies on a single crop, rather than on a more diversified system, will be more vulnerable to a pest affecting that crop. An area prone to water scarcity or a rainfed system will be more impacted by drought than an irrigated system. Therefore, small family farms entirely dependent on rainfed agriculture are more vulnerable from an economic point of view to drought than larger farms with other sources of water.

▶ **Poorer population groups:** Evidence suggests that, faced with a shock, poorer households are more likely to reduce consumption, while wealthier households have the capacity to access credit and savings and liquidate assets to cover current deficits.[265] This means choosing between limiting consumption and asset smoothing, with no other safer alternatives. Unsurprisingly, people from low-income groups are those most likely to migrate, but neither their capacity to cope and adapt to climate shocks nor their food security and nutrition necessarily improve when they move to urban environments. While year-round access to diverse and nutritious food may get better in urban areas for those who can afford it, reliance on highly processed, energy-dense foods and street foods tends to increase. The most affordable and available diets for poor urban populations are often unhealthy and adopting them could thus raise the risk of malnutrition and diet-related NCDs.

▶ **Populations groups that suffer greater inequality and marginalization:** There is mounting evidence of and broad consensus[266] on the inequality-driven impacts and risks related to climate shocks. Vulnerability emerges from the intersection of different inequalities and uneven power structures and is therefore socially differentiated.[267] For instance, the IPCC Fourth Assessment Report identified poor and marginalized indigenous peoples in North America[268] and in Africa[269] as highly vulnerable to climate shocks.

▶ **Women, children, the elderly and the socially isolated:** As highlighted before, vulnerability is often high for this group, which also includes indigenous and disabled people. These population groups experience multiple deprivations that inhibit them from managing daily risks and shocks[270] and present significant barriers to adaptation.

▶ **Men and women are impacted differently** by climate shocks. This difference arises from the distinct roles they have in society and from the way these roles are enhanced or constrained by other dimensions of inequality, risk perceptions and the nature of their response to hazards. As a result of extreme climate events and climate-related disasters, women often experience additional duties as labourers and caregivers due to, for instance, male out-migration. They face more psychological and emotional distress, reduced food intake, adverse mental health outcomes due to displacement and in some cases increasing incidences of domestic violence (Box 13).

▶ **Infants, young children and adolescent girls:** These young persons are often at higher risk and more vulnerable to climate variability and extremes due to more limited mobility, susceptibility to infectious diseases, reduced adequate care (including feeding and food intake) and social isolation. Adverse effects on the nutritional status in early life can irreversibly impair growth and development, »

BOX 13
THE GENDER DIMENSIONS OF VULNERABILITY TO CLIMATE SHOCKS

Women are particularly vulnerable to climate variability and extremes, and their vulnerability derives from restricted access to the social and environmental resources required for adaptation.

In many rural economies and resource-based livelihood systems, women have poorer access than men to financial resources, land, education, health, and other basic rights. Further drivers of gender inequality include social exclusion from decision-making processes and labour markets, making women less able to cope with and adapt to climate change impacts.[1]

In the Bongo district of north-eastern Ghana, households headed by men were found to be more resilient to climate shocks than those headed by women in terms of income and food access, assets and adaptive capacities. The reason for this inequality was women's limited rights in livelihood decision-making processes and access to land and other productive resources. Households headed by men were also found to adopt 0.8 times more adaptation measures than those headed by women.[2]

In most countries access to credit for female family farmers was found to be 5–10 percent lower than that of their male equivalents.[3] Moreover, social norms or time constraints may prevent women from seizing off-farm opportunities, which influences their level of vulnerability, incomes and ability to adjust their agricultural production. In some communities, only men have the right to cultivate certain crops or to access markets. In addition, many adaptation practices require investments in cash, time or labour and are thus costly for households with limited access to credit

and with few – mostly female – working-age adults.

In addition, as women act as primary caregivers and providers of food, water and fuel, they are more vulnerable when droughts and floods occur. For instance, with dry seasons now becoming longer, women are working harder to feed and care for their families without support. In central Africa, where up to 90 percent of Lake Chad has disappeared, women have to walk much further to collect water.

As an indirect social consequence of climate-related disasters, as well as slow-onset climate events, in Viet Nam[4] and Bangladesh[5] increased gender-based violence within households has been reported owing to greater stress and tension, loss and grief, and disrupted safety nets.

Finally, data from India indicate that exposure to a disaster (generally in the form of floods, droughts and extreme temperatures) had much worse effects on undernutrition among girls than boys, possibly because of differential parental behavioural responses and other disinvestments in girls' human capital.[6] In fact, persistent gender inequalities in nutrition may dictate that male children are prioritized over female children in intrafamily caregiving practices, food distribution and health care access and therefore have decreased risk of acute malnutrition. In Rwanda, girls born during crop failure showed stunted growth compared to those born when there was no crop failure (i.e. had 0.86 standard deviations lower height-for-age z-scores, with no adverse impacts noted among boys). The authors attributed the gender differences observed to preferential feeding of boys.[7]

SOURCES:

[1] J. Paavola. 2008. Livelihood, vulnerability and adaptation to climate change in Morogoro, Tanzania. *Environmental Science & Policy*, 11(7): 642–654; H. Djoudi and M. Brockhaus. 2011. Is adaptation to climate change gender neutral?: Lessons from communities dependent on livestock and forests in northern Mali. *International Forestry Review*, 13(2): 123–135; B. Rijkers and R. Costa. 2012. Gender and Rural Non-Farm Entrepreneurship. *World Development*, 40(12): 2411–2426.

[2] J.A. Tambo. 2016. Adaptation and resilience to climate change and variability in northeast Ghana. *International Journal of Disaster Risk Reduction*, 17: 85–94.

[3] FAO. 2011. *The State of Food and Agriculture 2010–11. Women in agriculture: closing the gender gap for development*. Rome.

[4] B. Campbell, S. Mitchell and M. Blackett. 2009: Responding to Climate Change in Vietnam. Opportunities for Improving Gender Equality. A Policy Discussion Paper. Ha Noi, Oxfam and UN.

[5] J. Pouliotte, B. Smit and L. Westerhoff. 2009. Adaptation and development: Livelihoods and climate change in Subarnabad, Bangladesh. *Climate and Development*, 1: 31–46; C. Stott. 2014. *An Examination of the Least Developed Countries in the IPCC AR5 WGII*. London, IIED.

[6] A. Datar, J. Liu, S. Linnemayr and C. Stecher. 2013. The impact of natural disasters on child health and investments in rural India. *Social Science & Medicine*, 76(1): 83–91.

[7] R. Akresh, P. Verwimp and P. Bundervoet. 2011. Civil War, Crop Failure, and Child Stunting in Rwanda. *Economic Development and Cultural Change*, 59(4): 777–810.

» school performance and earning potential throughout life. The impacts on their nutrition and health have already been discussed, but it is important to add further considerations. First, while adults and older children are more severely affected by some climate-sensitive vector-borne diseases such as dengue, young children are more likely to die from or be severely compromised by diarrhoeal diseases (caused by, for example, floods) and slide into the vicious cycle of infection and malnutrition. Second, as a consequence of climate extremes and climate-related disasters children may lose access to schooling and health care facilities and be compelled to work to support their families. This may put children and adolescent girls at increased risk of emotional, physical, and sexual violence.[271] Overall, climate shocks can thus exacerbate existing inequalities that disproportionality affect disadvantaged children and limit their opportunities for the future.

Policy and programme coherence is urgently needed to address the increased exposure and vulnerability of livelihoods, particularly of disadvantaged population groups. Without proper planning, climate variability and extremes will also affect vulnerability to future extreme events. Any rise in climate extremes can exacerbate the vulnerability of disadvantaged population groups with adverse long-term developmental effects if no action is taken to increase resilience at all levels (productive, social, climatic and environmental). ∎

2.4 WORKING TOWARDS COHERENCE OF POLICIES, PROGRAMMES AND PRACTICES TO ADDRESS CLIMATE VARIABILITY AND EXTREMES

KEY MESSAGES

➜ Scaled-up actions across sectors are needed to strengthen the resilience of livelihoods and food systems to climate variability and extremes. Such actions should take place through integrated disaster risk reduction and management and climate change adaptation policies, programmes and practices with short-, medium- and long-term vision.

➜ When designing policies and programmes it is important to consider that adaptation has limits in some contexts. This may necessitate the transformation of systems themselves in a manner that leads to increased resilience.

➜ Climate resilience is key and requires context-specific interventions aimed at anticipating, limiting, and adapting to the effects of climate variability and extremes and building the resilience of livelihoods, food systems and nutrition to climate shocks and stresses.

➜ To be successful across livelihoods and food systems and to address food insecurity and all forms of malnutrition, climate resilience policies and programmes should be built around climate risk assessments, science and interdisciplinary cross-sectoral knowledge, and participatory and inclusive blended humanitarian and development approaches driven by the needs of climate-vulnerable groups.

➜ Solutions require increased partnerships, enhanced risk management capacities and multi-year, predictable large-scale funding of disaster risk reduction and management and climate change adaption policies, programmes and practices.

→ Implementation of climate resilience policies and programmes means adopting and refitting tools and interventions such as: risk monitoring and early warning systems; emergency preparedness and response; vulnerability reduction measures; shock-responsive social protection, risk transfers and forecast-based financing; and strong risk governance structures in the environment–food–health system nexus.

The analysis and evidence presented up until now shows how climate variability and extremes are undermining food availability, access, utilization and stability. They are also challenging health and caregiving practices and are thus among the underlying causes of food insecurity and malnutrition in several parts of the world. The critical aspect going forward is to build lasting climate resilience, which will require scaled-up policies, programmes and practices and better ways of working to ensure their success.

Resilience is generally understood as the capacity of individuals, groups, communities and institutions to anticipate, absorb (i.e. cope), adapt and transform in the face of shocks.[272] Interventions aimed at reducing vulnerabilities and enhancing resilience should therefore look to strengthen these capacities in anticipation of and in reaction to climate variability and extremes that undermine food security and nutrition.

The concept of resilience, and more specifically climate resilience, plays an important role in global policy processes. This section describes the existing global policy frameworks and concepts that can provide the basis for efforts in building climate resilience, as well as the need to reduce the fragmentation of the interventions among global entities and partners. This section also notes that while national and local governments can be guided by better integrated global policy processes, they also need to overcome a number of context-specific challenges when trying to determine measures to prevent risk and address the effects of increased climate variability and extremes. In view of the challenges at all levels (global, national and local) and the complexity involved in building climate resilience, the section also provides recommendations on the cross-cutting factors and the specific tools and mechanisms that can

lead to successful policies and practices that address climate risks.

Global policy frameworks, processes and concepts for addressing the threats and impacts of climate variability and extremes on food security and nutrition

It is important to understand a number of global policy dimensions and a wide range of different actors when examining the possible solutions for addressing the threats and impact of climate variability and extremes on food security and nutrition. Four United Nations frameworks and a multi-stakeholder global process are particularly important (Figure 40).[273] Each provides key concepts, though in rather siloed policy areas, with different platforms and processes, involving government and other stakeholders and technical experts:

▶ **The United Nations Framework Convention on Climate Change (UNFCCC)** – through which **the 2015 Paris Agreement** was negotiated – offers the policy architecture to support climate change adaptation and mitigation goals. **Climate change adaptation (CCA)** comprises actions to manage and reduce the risks and impacts of climate-related hazards, climate variability and gradual climate change at large. Nationally Determined Contributions (NDCs),[274] National Adaptation Plans (NAPs)[275] and National Adaptation Programmes of Action (NAPAs)[276] reflect countries' CCA.

▶ **The Sendai Framework for Disaster Risk Reduction (SFDRR)** (2015–2030), adopted in 2015, provides a worldwide framing for disaster risk reduction (DRR) and disaster risk management (DRM) work, which includes humanitarian disaster management or emergency response. Disaster risk management is considered to be the application of DRR policies and strategies in the cycle before, during and after disasters.[277] DRR and DRM are rooted in the humanitarian and development fields and are supported globally by **the United Nations Office for Disaster Risk Reduction (UNISDR)**. Focusing on extreme events and combining both immediate disaster management and longer-term risk

FIGURE 40
GLOBAL POLICY PLATFORMS AND PROCESSES WHERE CLIMATE RESILIENCE IS A KEY ELEMENT FOR THE ACHIEVEMENT OF SUSTAINABLE DEVELOPMENT

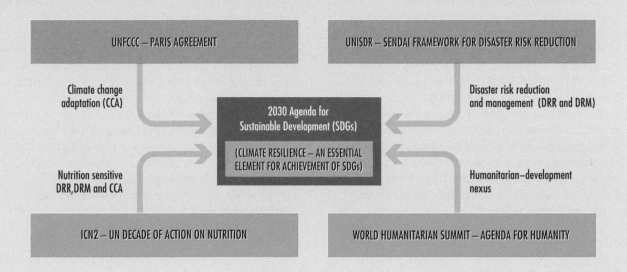

SOURCE: FAO.

prevention, DRR outlines policy objectives and the strategic and instrumental measures employed to anticipate and prevent future disaster risk in order to reduce existing vulnerability and exposure in the face of hazards, including climate extremes.

▶ The global ambition of "Transforming Our World: **the 2030 Agenda for Sustainable Development**" was adopted by world leaders during the 2015 United Nations Summit. This global policy framework commits the international community to end poverty, hunger and malnutrition, tackle climate change and achieve equitable and sustainable development in its three dimensions (social, economic and environmental) by 2030.[278] Achieving the 17 Sustainable Development Goals (SDGs) of the agenda calls for comprehensive, coherent, convergent and participatory approaches from all stakeholders, including humanitarian, development, peace and climate actors. Many SDGs – in particular SDG1 on ending poverty, SDG2 on ending hunger and SDG13 on combating climate change – have specific targets on resilience.[279]

▶ At **the Second International Conference on Nutrition (ICN2) in 2014**, countries committed and adopted the global policy framework to end all forms of malnutrition – in **the Rome Declaration on Nutrition and the Framework for Action,** fostered by the proclamation by the UN General Assembly of the United Nations Decade of Action on Nutrition, 2016-2025. The ICN2 outcomes recognize the need to address the impacts of climate change and to enhance the resilience of the food supply in crisis prone areas. The United Nations Decade of Action on Nutrition provides an operational framework for strengthening efforts to end hunger and eradicate all forms of malnutrition worldwide, including through nutrition sensitive disaster risk reduction and climate adaptation policies and programmes to strengthen the resilience of people's livelihoods and food systems for healthy diets.

▶ **The World Humanitarian Summit and the Grand Bargain,**[280] held in 2016 in Istanbul and known as the Agenda for Humanity, is a multi-stakeholder global policy process that pursues three goals: to reinspire and reinvigorate a commitment to humanity and the universality of humanitarian principles; to initiate a set of concrete actions and commitments aimed at enabling countries and communities to better prepare for and respond to crises, and be resilient to shocks; and to share best practices that can help save lives around the world, put affected people at the centre of humanitarian action and alleviate suffering. The resulting multi-stakeholder commitments for action revolve around five core themes including leaving no one behind and working differently to end needs. These themes include a work stream on risk and vulnerability reduction with a focus on natural hazards and climate change where resilience is key.[281]

Though these global policy frameworks and processes lack alignment they all include the objectives of eradicating hunger and malnutrition, reducing poverty and addressing the underlying root causes of vulnerability for building resilience against multiple risks, including those associated with the climate. In addition, they call for a transformative shift to put the world onto a more resilient and sustainable pathway.

Today, the much needed convergence and coherence of climate resilience actions by humanitarian and development actors is being promoted through another important dialogue called the humanitarian–development nexus. This considers how to bridge the needs of people across the current artificial divide between humanitarian and development responses, incorporating the concept of resilience along the continuum. While lacking the more formal policy architecture of CCA and DRR, the nexus debate was re-energized during the World Humanitarian Summit in 2016. More recently the humanitarian–development nexus has also incorporated peace considerations – otherwise known as the triple nexus – aligning it even closer to the 2030 Agenda.

Existing challenges for countries in responding to climate variability and extreme events

National and local governments are facing a number of challenges in trying to determine

measures to prevent risk and address the effects of increased climate variability and extremes.

To begin with, each of the global policy platforms compartmentalizes different concepts and expertise into silos of action across and within sectors. This introduces potential inefficiencies in overlapping interventions and missed opportunities for integrating responses, while also diluting available funds and human resources. Integration and convergence of efforts are critical for addressing climate risks in general, but even more so for bringing together food systems, agricultural livelihoods, and food security and nutrition and for promoting sustainable, healthy diets as part of climate resilience action plans.[282]

For adaptation action, National Adaptation Plans (NAPs) and Nationally Determined Contributions (NDCs) could be a prime instrument for implementation. Almost 90 percent of developing countries have designated the agricultural sector as a priority for adaptation actions in their NDC[283] and a similar prioritization is found in DRR plans. However, poorly defined institutional roles between different ministries and capacity gaps – as well as compartmentalized approaches and actions on agriculture (including crop, livestock, fisheries, aquaculture, and forestry sub-sectors), food security, nutrition and health – are hindering integrated DRR/DRM and CCA policies, programmes and practices for resilience. Furthermore, less than 1.5 percent of international financing for climate change adaptation is currently allocated to health projects.[284]

Another challenge is that adaptation has limits, a critical aspect to keep in mind when designing measures to prevent risk and address the effects of increased climate variability and extremes. Agricultural crops, fish and seafood species, coral reef and forest ecosystems, and even human beings are all constrained by climate thresholds.[285] Adaptation is no longer feasible once these thresholds are reached, and the implications of this are significant. For example, lack of possibilities for adaptation is the very reason why the likelihood of a person being displaced by a disaster is 60 percent higher today than it was four decades ago.[286]

In some cases policy design also needs to recognize the possibility that the limits to adaptation may force people to transform or change their system of reference.[287] For example, small family farms that are faced with unreliable agriculture productivity (due to climate variability and extremes) may only find solutions to this problem by rethinking their entire livelihood system. Policies must also ensure that such changes ultimately help increase resilience. Migration is an example of a transformational adaptation strategy that may not necessarily increase resilience.

Evaluating the suitability of scaling up tested DRR/DRM and CCA options in some locations can be seriously hampered by a lack of technical capacity and data. Inadequately understanding and measuring how climate variability and extremes affect livelihoods and food systems in different contexts often leads to the design and development of policies and plans that do not contribute to resilience building.[288] This is further complicated by the comprehensiveness of food systems and the interrelated nature of climate, food systems, livelihood systems, nutrition and health issues.[289]

There are still challenges related to data collection and management to assess and better understand losses and damages linked to climate variability and extremes. The absence of well-defined or well-established indicators and monitoring and evaluation systems remains problematic due to the range of conceptual frameworks and institutions involved across this spectrum of work. Addressing these gaps is fundamental not only for ensuring well-tailored policies and investments but also for tracking progress toward global targets related to the SFDRR, the Paris Agreement and the SDGs.[290]

On a more positive note, the growing focus on building resilience – and specifically climate resilience – which incorporates the concept of climate risk management, is helping to create a bridge between DRR/DRM and CCA and is providing important guidance to stakeholders for integrating these concepts into policies, programmes and actions. In 2017 a number of high-level international meetings began to promote integrated approaches with a focus on climate resilience, including: the UNFCCC's Subsidiary Body of Scientific and Technological Advice (SBSTA); the Global Platform for Disaster Risk Reduction; and initiatives such as the UN Climate Resilience Initiative (A2R)[291] and the Capacity for Disaster Reduction Initiative (CADRI) global partnership. These efforts towards integration and coherence with greater focus on resilience will hopefully lead to enhanced, coordinated and coherent sectoral policies, investments and programmes, as well as more effective and holistic actions for climate resilience of the agriculture, food security and nutrition sectors.

To meet the needs of the most vulnerable groups, cross-institutional partnerships, responsibility sharing and information flow need to be at the centre of an inclusive climate resilience strategy within and across sectors. While the 2030 Agenda recognizes this need, more efforts are required at national and local level. Resilience building must be realized through nutrition-sensitive measures blending short-, medium- and long-term interventions that link humanitarian disaster response and risk-informed development actions addressing root causes of climate vulnerabilities and CCA. Longer-term strategies designed to increase general food system resilience will improve food security and nutrition for present and future generations.[292]

Cross-cutting factors that lead to successful policies and practices addressing climate risks

Designers of policies, programmes and practices need to be mindful of the key elements that determine their success or failure. Climate risk assessments are fundamental for understanding risks and impacts across agriculture, food security and nutrition sectors in order to adequately evaluate options and inform decision-making. Science is critical for identifying appropriate solutions, including technological ones. Participatory, inclusive and equitable gender-based approaches must guide the entire policy/programme cycle, putting vulnerable groups at the centre of responses. The comprehensiveness of the food system needs to be understood, including how it can be transformed to address climate-risk, environmental, nutrition and health-sensitive considerations. Transformational change will not happen without dependable, multi-year and large-scale financing as well as shared climate resilience good practices and knowledge management.

Climate risk assessment at the core of policy, programme and practice design

Policies, programmes and practices are ineffective if they do not help individuals anticipate, absorb and adapt to climate variability and extremes. Their design and monitoring thus requires comprehensive risk assessments and an understanding of the potential impacts of climate variability and extremes on human, natural and food systems. Assessments should be quantitative to a large extent, because policy-makers need to have a sense of the magnitude of impacts and the measures to offset them; but they also need qualitative insight. A number of methodological tools are available so that these assessments are implemented with climatic, biophysical and economic orientations and stakeholder engagement, as well as a focus on the impacts on agriculture, livelihoods, nutrition, health, resilience, poverty and inequality.[293]

Assessments also need to be based on risk-specific and local contexts, with an understanding of how livelihoods, food security, nutrition and food systems are affected and interconnected. This is essential to better differentiate between affected groups, identify their specific needs, including gender and target them with shock-specific and context-relevant programme options and measures to enhance resilience. The critical aspect is that assessments produce people-centred results that inform decision-making.

In Sri Lanka – a country with high exposure to climate extremes (see Annex 2) – WFP and FAO have been working with the government, farmers and other vulnerable groups to identify the best strategies to improve climate resilience, sustainability and self-sufficiency.[294] Climate risk analyses show that any intervention should consider longer-term projections for sea-level rise and salt intrusion, as current interventions – in areas where levels of food insecurity and undernutrition are high – do not necessarily align with future climate risks.[295]

Integrating climate information in socio-economic and environmental analysis is critical for understanding current trends and for targeting risk reduction and adaptation measures towards the most vulnerable groups in the most vulnerable areas. Studies for Malawi and Zambia highlight that different types of exposure to climate risk call for different types of adaptation strategies.[296] Not every farmer will benefit from the same adaptation strategy in a risk-prone area. In Zambia poor households can reap significant benefits from adopting crop diversification strategies, whereas diversification may not be particularly beneficial for wealthier households whose returns to specialization are high.[297]

Cost–benefit analysis (CBA) can help policy-makers explore alternative options and expected net benefits in order to determine the best allocation of resources.[298] For example, CBA has been used to evaluate investment options in National Adaptation Plans.[299] In Kenya, Zambia and Uruguay, ongoing CBA studies containing climate scenarios have been used within the Integrating Agriculture in National Adaptation Plans (NAP-Ag) Programme.[300] One lesson learned from these experiences is that CBA analysis should be complemented with qualitative assessments of both barriers to adoption as well as environmental and social impacts of adaptation strategies.

Science and interdisciplinary knowledge to inform technological solutions

Technological solutions that farmers may adopt will also have to be informed by climate-related science and evidence. Scientific climate information is key to enhancing the accuracy and the role of preparedness and adaption mechanisms, such as forecast-based financing mechanisms, weather-based index insurance and shock-responsive social protection, among others. It is important to develop accurate climate and weather forecasts to design triggers for the quick dispersal of finances or the provision of safety nets to those affected – or likely to be affected – by a climate event.

New sources of knowledge beyond formal research systems that include local indigenous knowledge are also critical for agricultural innovation systems.[301] For example, research

BOX 14
ENHANCING THE CONTRIBUTION OF NEGLECTED AND UNDERUTILIZED SPECIES (NUS) TO FOOD SECURITY AND INCOME

Neglected and underutilized species (NUS) constitute a large portfolio of plant genetic resources that include cultivated, semi-domesticated or wild species not treated as commodities. They are cultivated by rural communities according to traditional knowledge and practices, using low-cost inputs. Because NUS occupy important niches and are adapted to local conditions, they serve as a safety net for indigenous farmers whenever staple crops fail during periods of stress or following disasters. As they are often bred by breeders, researched by agricultural scientists and promoted by policy-makers, they could make substantial contributions to income generation, resilience and adaption to climate change among small-scale family farmers.

In the Andean region of South America, research conducted by Bioversity International, and supported by IFAD, worked with three types of NUS crops, namely: Andean grains (such as quinoa and amaranth); minor millets (such as finger millet, little millet and barnyard millet); and medicinal and aromatic plants (such as argel, caper, oregano and mint). These were used to test innovative approaches to sustainable conservation and cultivation by incorporating local indigenous knowledge, and to inform related research work on climate change and its impact on local food production systems.

Using local indigenous knowledge and merging it with novel cultivation practices, small family farm households who cultivate NUS crops could benefit from stronger food production systems, which can improve food security, increase income-generating opportunities, and enhance coping mechanisms against climate change.

SOURCES: S. Padulosi, N. Bergamini and T. Lawrence, eds. 2012. *On farm conservation of neglected and underutilized species: status, trends and novel approaches to cope with climate change.* Proceedings of an International Conference, Frankfurt, 14–16 June 2011. Rome, Bioversity International; S. Padulosi, J. Thompson and P. Rudebjer. 2013. *Fighting poverty, hunger and malnutrition with neglected and underutilized species (NUS): needs, challenges and the way forward.* Rome, Bioversity International.

conducted by Bioversity International demonstrated that neglected and underutilized species (NUS) can help contribute to increasing food security, income and resilience to climate change, as illustrated in Box 14.

The successful inclusion of local indigenous knowledge into research for agricultural innovation systems requires an interdisciplinary effort under the wider banner of climate services, involving meteorologists, agronomists, nutritionists, communications specialists, development practitioners and communities themselves in the co-production of climate information tailored to meet stakeholders' needs.[302] It is important to identify the right communication channels so that people can easily access this information and make appropriate decisions.

Such interdisciplinary or cross-sectoral efforts are found in the climate-smart agriculture (CSA) approach, which requires site-specific assessments to identify suitable agricultural production technologies and practices for specific climate-related shocks and stresses in a given location. This approach allows for weaving together risk mitigation and climate change adaptation, by focusing on three pillars: (i) increasing agricultural productivity and income; (ii) strengthening resilience and adaptation; and (iii) reducing and/or eliminating GHG emissions. CSA focuses on developing the technical, policy and investment conditions to achieve resilient and sustainable agricultural development for food security and nutrition in the face of climate change.[303] It also assesses the interactions between sectors and the needs of different involved stakeholders.[304]

There are certain well-known, site-specific climate-smart technological solutions that have been tested and are already supporting climate resilience building. These include new crop varieties and livestock breeds; efficient water management (including new water sources, irrigation, drainage, water harvesting and saving technologies, desalinization and storm- and wastewater management); conservation agriculture; climate-proof food storage and preservation facilities; flood and cyclone shelters; and climate risksensitive infrastructures. Deploying these solutions requires analysing and identifying climate risks and impacts as well as costs, benefits, incentives and barriers to their adoption. Many of these climate-smart technological solutions also help reduce greenhouse gas (GHG) emissions.[305]

Adapting and reducing GHG emissions through a climate-resilient food systems approach broadens the range of opportunities and facilitates consideration of systems-level effects and interactions. It is critical to go beyond a focus on agriculture and production to consider the interlinked nature of livelihoods and food systems and the implications for building climate resilience as part of a wide-ranging transformation of food systems for improved nutrition and sustainable healthy diets. In Malawi, for example, crop diversification is an important adaptation strategy that – when implemented with a food systems approach – can benefit food security, health and nutrition while helping reduce the vulnerability of small-scale family farmers to income volatility resulting from climate variability and extremes (see Box 15).

Knowledge generation and sharing with regard to resilience good practices

Systematic documentation of good practices for climate resilience should be planned at the outset of the design of any intervention. Indicators should be defined not only to monitor and evaluate impact but also to capture the process of implementation in order to understand why some solutions work over others. Knowledge management platforms are a valuable vehicle for countries and communities within countries to share lessons, experiences and good practices and to support each other in accelerating implementation of relevant, context-specific climate resilience actions. It is worth noting that solutions that specifically address climate

risks and shocks are not only stress-specific, but also sector-specific and essentially site-specific, meaning that replicating interventions in different contexts requires a careful examination of how better to contextualize the practices to respond to the specificities of each context.

More efforts are needed in making information and good practices on climate resilience accessible to most vulnerable households and communities. This includes establishing knowledge-sharing mechanisms that enable people to participate in the design of context-relevant interventions to enhance climate resilience. Novel ways of sharing information with communities include participatory videos, which have proven effective in spreading knowledge of successful climate adaptation practices with others.[306]

Participatory approaches for local solutions

Supporting climate resilience-building efforts requires site-specific solutions that are owned by the communities that they intend to help. A participatory, inclusive, equitable and gender-based approach is critical to bringing local stakeholders together to identify needs through a better understanding of the climate vulnerabilities and risks faced by communities and individuals. Likewise, it is important to take advantage of autonomous (i.e. local) knowledge and practices when addressing climate variability and extremes. Engaging local people and encouraging open community consultation when designing and implementing interventions helps to build community ownership and ensure long-term sustainability, while also taking into account cultural and gender issues.

A range of locally appropriate climate-resilient options should be designed and implemented through inclusive and gender-sensitive participatory processes. These should be present throughout, beginning with the initial vulnerability and risk analysis, continuing through the prioritization of choices and moving forward to the implementation of measures, taking into account the availability of local resources and the anticipated costs and benefits in the short and long term.[307] It is important to maintain community engagement throughout »

BOX 15
CLIMATE-SMART AGRICULTURAL PRACTICES AND FOOD SYSTEMS: THE CASE OF SMALL FAMILY FARM CROP DIVERSIFICATION IN MALAWI

In sub-Saharan Africa, many countries' national food security relies on a few staple crops, particularly maize. This crop is produced mostly by small-scale family farmers under rainfed conditions, which makes households and national food security vulnerable to climate variability and extremes.

As seen in this report, climate variability and extremes can negatively impact on small family farm incomes as agricultural production falls. For some Malawian households food consumption declines not only because of decreases in income but also because households have less of their own food production to consume.

Crop diversification is an important adaptation and vulnerability reduction strategy that can, in the context of increased climate variability and extremes, help distribute risk, increase productivity and stabilize incomes of small-scale family farmers, thus

improving food access. In Malawi, more diversified cropping systems – particularly those that incorporate legumes – have been shown to significantly reduce crop income variability compared with maize monocropping (see figure below).

Through crop diversification, farming households can spread production and income risk over a wider range of crops. Moreover, diversification can produce agronomic benefits in terms of pest management and soil quality and nutritional benefits by promoting dietary diversity depending on the crop combination.

Though crop diversification can be an important adaptation and risk reduction strategy, to achieve climate resilience it needs to be implemented with a food systems approach that ensures functional and competitive private input and output markets, and addresses other key interlinked factors in the food systems.

CROP DIVERSIFICATION REDUCES INCOME VOLATILITY

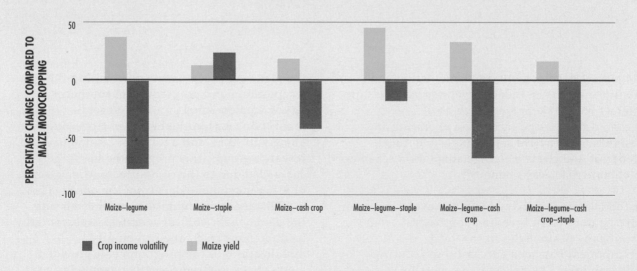

SOURCE: FAO, Economic and Policy Analysis of Climate Change (EPIC) Team of the Agriculture Development Economics Division (ESA).

SOURCES: FAO. 2018. *Crop diversification increases productivity and stabilizes income of smallholders.* Rome; FAO. 2016. *Managing climate risk using climate-smart agriculture.* Rome.

BOX 16
PARTICIPATORY PLANT BREEDING TO INCREASE CROP YIELDS AND RESILIENCE IN IRAN (ISLAMIC REPUBLIC OF)

Losing agricultural biodiversity reduces the opportunities to cope with future challenges, including a changing climate. Biodiversity is also an important driver for enhancing the resilience of small-scale family farmers to climate change, drought, and pest and disease outbreaks, among other things. In the Islamic Republic of Iran, planting only a small number of improved crop varieties in place of a mix of several traditional varieties has resulted in the loss of genetic diversity in agricultural systems. Thus, farmers need seeds that are better-adapted to increased climate variability and other climate shocks.

Traditional crop varieties represent a valuable source of increased agricultural diversity as they have evolved through a combination of adaptation to local environments and generations of genetic selection. It is widely recognized that traditional varieties often have much greater resilience to drought and other stresses, although they yield lower outputs in favourable conditions. Furthermore, they often do not need chemical pesticides and fertilizers and require less water.

The Centre for Sustainable Development (CENESTA), the International Center for Agricultural Research in the Dry Areas (ICARDA), the Rice Research Institute of Iran, the departments of agriculture in project provinces, farmers and farm associations, along with IFAD, introduced the concept of evolutionary participatory plant breeding with the aim of increasing crop yields and resiliency using site-specific approaches. In this approach, farmers used the best seeds from field trial plots combined with traditional varieties for the next planting season to create a mix of varieties that were highly regulated. After only one cultivation season, this approach yielded greater results than cultivating only a single crop variety. By growing this mixture of crop varieties, the crops became more climate-resilient: the increased diversity of their genes allowed them to evolve and adapt to climate variability and unpredictable weather patterns.

SOURCES: Centre for Sustainable Development and Environment (CENESTA). 2012. *Evolutionary Plant Breeding: Guide for farmers-facilitators.* Tehran; R. Pilu and G. Gavazzi. 2017. *More Food: Road to Survival.* Sharjah, UAE, Bentham Science Publishers.

» the project development, implementation and monitoring phases. Indeed, even researchers now interact with stakeholders such as decision-makers and farmers in exploring and designing alternative sets of plausible future scenarios and climate change adaptation schemes in climate risk assessments.[308]

In the Islamic Republic of Iran, an evolutionary participatory plant breeding approach specifically designed to fit the local environment has been successful in reducing vulnerabilities of small-scale farmers by improving crop yields as well as increasing crop resilience to drought and other stressful events, as illustrated in Box 16.

An example of where participatory approaches have been particularly successful can be seen in the planning process embedded within the Adaptation Fund programme in southern Egypt. Implemented jointly by several institutions within the Egyptian Government, along with WFP and a range of community and research groups, this programme has largely succeeded due to the committed participation of different stakeholders from the outset. The establishment of committees at all levels and the deployment of local volunteers substantially facilitated outreach and community mobilization for the programme. It provided alerts for two extreme weather events in the 2013 and 2015 seasons along with recommendations on how to reduce losses.[309] In 2016 and 2017, this same early warning system helped farmers of wheat, sorghum and maize reduce their losses from heatwaves by around 70 percent.

Empowerment of women and vulnerable groups

Building resilience to climate variability and extremes requires gender-sensitive policies, planning, budgets, technologies, practices and processes accessible to both men and women farmers. Although women comprise, on average, 43 percent of the agricultural labour force in developing countries and are key for food security and nutrition, they generally have less access than men to productive resources and opportunities.[310] Building resilience thus requires a solid understanding of gender-based differences and interventions that are risk- and gender-responsive. The R4 Rural Resilience Initiative (R4), launched by WFP and Oxfam America in 2011,[311] shows the benefits of gender-responsive programming in Ethiopia, Malawi, Senegal and Zambia, where women are becoming less vulnerable to climate risks and more empowered to support themselves and the food security and nutrition of their families. This is due to an integrated package of financial services and community assets used to address climate variability and extremes. An impact evaluation found that households headed by women in Ethiopia had the largest gains in productivity and farm investments and faced fewer climate-related food shortages.[312]

In capture fisheries, particularly in sub-Saharan Africa, women are predominantly engaged in processing, trading and selling. The estimated 56 million women in this sector are constrained by deplorable working conditions, poor market and transportation infrastructure, limited financial and business services, competition for limited catches, and variable supply. Investments that strengthen women's empowerment in this sector have been proven to lead to improvements in nutrition and health of women and their families.[313]

The needs of other vulnerable groups should also be at the forefront of policy responses. Infants and young children are particularly vulnerable to climate shocks, which can diminish their food security and nutrition, thus limiting their future opportunities. Children are notably affected if, for example, such shocks undermine their school performance, decrease their earning potential, or expose them to a higher risk of diet-related non-communicable diseases later in life. Further, the poor nutritional status of pregnant women or adolescent girls who are more exposed to climate impacts increases the risk that their children will suffer from poor health outcomes.[314]

Acknowledging these risks to nutrition from changing climate variability and extremes is critical in creating more effective safety nets or social protection schemes that are responsive to climate risk.[315] Interventions should also consider advocacy across all agencies and actors in the public, private and civil society sectors to protect and build coping and adaptation strategies for women and other vulnerable groups.

Integration of interventions to enhance climate resilience of the entire food system

The more integrated sets of interventions are within and across sectors, the better they are in meeting household, community and institutional needs in the face of climate variability and extremes. Coordination is a prerequisite in ensuring people and institutions work across all agriculture sectors as well as other sector such as health, eduction, water and energy. This is particularly the case for enhancing the climate resilience of the food system as a whole, thus contributing to healthy diets for all. Nevertheless, while there is immense potential for synergies, the potential trade-offs also need to be considered.

Much of the work on DRR and CCA relies on strengthening climate risk assessment capacities within and across sectors. The ICN2 Framework for Action is meant for use by governments and other stakeholders to guide cross-sectoral implementation. The Work Programme of the UN Decade of Action on Nutrition emphasizes priority actions in areas that are entry points to integrating climate change and food security questions into health risk assessments. These include sustainable, resilient food systems for healthy diets and safe and supportive environments for nutrition at all ages. This points to a unique opportunity to address the challenge of existing fragmented global policy processes and the need to forge synergies for better dialogue among climate, humanitarian, development, nutrition and health actors in the spirit of the universal SDGs.

The interconnected nature of DRR/DRM, CCA and the environment–food–health nexus means that there is potential for scaling up solutions that generate co-benefits for the environment, climate, nutrition and health. This nexus requires political dialogue and advocacy to enhance social participation and engagement of a wide range of actors, including environmental groups, consumer and health advocates, health care providers, farmers and farm workers, large and small private sector entities and citizens.

Given that climate variability and extremes are affecting the quantity, diversity and quality of food available and consumed – thus potentially undermining nutrition – healthy diets need urgent protection. Climate-smart technologies can support food diversification, incentivize the production of more nutrient-dense foods, reduce the impact of climate-related stresses on crop and livestock quality, and, more broadly, help improve the efficiency and resilience of the food system.

Integrating climate and food security questions into health risk assessments is also important in providing early signals for potential outbreaks of disease, thus triggering early action. There are significant benefits to coordinating needs assessments in livelihoods, nutrition, health and other sectors to save more lives and protect and restore more livelihoods.[316] Such assessments are already identified in key humanitarian indicators for country teams to create a composite, ongoing picture of emergencies.[317]

Nutrition-sensitive and risk-responsive social protection programmes can also safeguard nutrition before and during climate shocks, especially if they allow households or nutritionally vulnerable groups – such as young children and pregnant and lactating mothers – to be able to afford nutrient-rich locally produced foods and maintain dietary diversity before, during and after a shock. Climate risk strategies need to include local diet quality goals, which can be achieved when there is a better understanding of: how longer-term climate change will affect the suitability of local crops in a specific site; whether access to fresh fruit and vegetables, meat and dairy products will be disrupted;[318] and what new agricultural and

livelihood practices avoid jeopardizing people's basic nutritional food basket.[319]

However, safeguarding nutrition before or when climate variability and extremes strike will have to go hand in hand with a variety of risk reduction and adaptation options that governments and communities can implement to protect public health. As part of efforts to ensure universal health coverage, measures to strengthen the resilience of health systems in order to manage climate risks include: considering partnerships between DRR actors, NGOs, the private sector (while avoiding conflict of interest) and national health systems within DRR plans; enhancing early warning systems and emergency preparedness for rapid response and recovery from extreme climate events; and protecting critical health infrastructure from extreme climate events.[320] It is important to have stronger surveillance systems in place that can identify food safety issues and infectious diseases, so that control systems can rapidly and accurately notify populations at local, national and international levels.

Furthermore, investment in universal health coverage that both ensures primary health care interventions and builds community resilience is key. Funding needs to address the determinants of environmental and social health (such as housing safety and air, water and food quality) under various climate conditions; improve social welfare in emergency situations; and provide essential nutrition actions,[321] including screening for and managing cases of child and adult malnutrition. It is equally important to take into account the diverse composition of modern communities (including migrants and different ethnicities) as well as differences in health-seeking behaviours.

Dependable, multi-year and large-scale funding streams and mechanisms

Integration of short-, medium- and long-term interventions and actors to achieve climate resilience also requires dependable, multi-year and large-scale funding. Evidence shows that lack of funding has resulted in a decline in development gains following the impact of climate extremes and variability. Humanitarian responses present many examples where slow-onset climate shocks have been identified well ahead of time, but a lack of early-action funding resulted in devastating impacts on people. The most salient recent example is the famine in Somalia following a drought across the Horn of Africa and the negative impacts for many food-insecure populations during and after the El Niño event in 2015–2016.

Responses to humanitarian crises – those arising from a combination of climate variability and extremes and political and social fracturing, among other factors – have cost a multiple of what they would have if investments had been made at an early stage when the crises were known to be developing.[322] This makes a clear economic argument for investment in multi-year resilience programming. The net cost of late response is estimated to be five to seven times higher than multi-year resilience building.[323] A study of WFP's response to the 2004–2005 food crisis in the Niger found that the cost of assistance to a person ten months after the initial appeal was three times the cost at just four months – a threefold increase in costs within a six-month period.[324]

Financial constraints have limited much Early Warning/Early Action (EWEA) to date, with dire implications for food security and nutrition. An ODI study highlights the weakness of DRR financing to drought-affected countries across two decades in Eritrea, Kenya, Malawi, the Niger and Zimbabwe, where over 100 million people were affected by drought, but their combined DRR financing was only USD 116 million.[325] Evidence from the 1998 floods in Bangladesh demonstrated nutritional impacts resulting from worsening food access and caregiving practices, thus increasing children's exposure to contaminants and malnutrition.[326] During the same crisis, government programme responses before the floods were shown to be more effective in protecting the well-being of children than responses after the floods.[327] Gaps in climate funding have also been outlined by the UNFCCC, which estimates the need for adaptation finance in developing countries at between USD 28 billion and 67 billion by 2030 – far outweighing current available funds.[328]

Overcoming these financial constraints is important for successful programmes to achieve scale, as vulnerability reduction measures across and within sectors are priorities that particularly require largescale financing (Box 17). Climate resilience programmes generally need dependable multi-year funding to succeed and show evidence of results in order to obtain further investment.

Specific tools and interventions that address climate risks

The following set of tools and interventions is based on approaches promoted by the Sendai Framework for DRR (SFDRR) that have been adopted and adapted to address climate risks that pose challenges to people's livelihoods, food security and nutrition. They typically encompass the cross-cutting features outlined above but deserve particular attention here to highlight how they can significantly contribute to building climate resilience.

Climate risk monitoring and early warning systems

Climate risk monitoring and early warning systems are among the most well-known tools available to governments and international agencies. They can prove to be essential in monitoring multiple hazards – and climate hazards more specifically – and in predicting the likelihood of climate risks to livelihoods, food security and nutrition. They are particularly useful when timely alerts help trigger accurate decision-making and early actions at all institutional levels, including in communities.

Early Warning Early Action (EWEA) systems focus on consolidating available forecasting information and triggers that put in place preparative and early actions to reduce the impact of a range of different hazards, including climate extremes.[329] Knowledge-sharing mechanisms for vulnerability reduction are also important in preparing both decision-makers and communities to implement early actions for projected shocks and changes.[330]

In anticipation of 2015–2016 El Niño impacts, WFP used seasonal climate forecasts to trigger early action in vulnerable communities in Zimbabwe. It promoted the cultivation of drought-tolerant small

BOX 17
INVESTING IN VULNERABILITY REDUCTION MEASURES, INCLUDING CLIMATE-PROOF INFRASTRUCTURE AND NATURE-BASED SOLUTIONS

Investments in vulnerability reduction measures (in line with DRR prevention and impact reduction actions, as per the Sendai Framework for DRR) need to be stepped up dramatically across and within sectors. These measures – also known as climate change adaptation and resilience measures (in line with Paris Agreement language) or simply CSA – include climate-resilient good practices at farm levels as well as climate-proof infrastructure and nature-based solutions.

Examples of vulnerability reduction measures against climate variability and extremes have already been presented in Box 14, 15 and 16 and others are found in Box 18. These shock-, sector- and context- or site-specific measures include: the use of adapted quality seed varieties and conservation agriculture for the crop sector; the improvement of resilient livestock breeds; the building of water points and cisterns for improved water management and conservation; and agroforestry and coastal mangrove protection and management. There is a wealth of documented climate-resilient good practices for agriculture, food security and nutrition; some of these are available on the Knowledge Sharing Platform on Resilience (KORE)[1] or other places.

Vulnerability reduction measures also include the implementation of Nature-based Solutions (NbS) as reflected in the outcome document of the COP23 high-level event on resilience[2] where it was stressed that healthy and diverse agro-marine ecosystems play a double role for a climate-resilient planet, as they: (i) buffer the impact of climate hazards such as drought, floods and storms and sea level rises; and (ii) provide essential ecosystem services such as fresh water, clean air, fertile soil, pollination and biodiversity, which contribute to fighting hunger and building resilient livelihoods, and are also crucial to sustaining the food system and life as a whole.

Working with nature involves implementing actions to protect, sustainably manage and restore natural or modified agromarine ecosystems. These systems simultaneously provide defence and life support benefits, including water and food for the poor and for the rich across borders, thus reducing food insecurity and poverty and enhancing climate-resilient livelihoods and food systems at large.

Climate-resilient and sustainable agricultural livelihoods are possible and can yield mitigation, adaptation and resilience co-benefits.

Around the world, it is essential to support countries in sustainably increasing their agricultural productivity while at the same time reducing climate risks. For example, the Rome-Based Agencies (RBA) work in the Corredor Seco in Central America region to increase the resilience of small-scale producers through ecosystems management and risk-informed, agro-ecological good practices.

Building climate resilience by working with nature implies reshaping investments at scale in healthy and diverse terrestrial and marine ecosystems that perform disaster risk reduction and climate change adaptation functions and are central for securing productive food systems and fighting hunger.

SOURCES:
[1] www.fao.org/in-action/kore/en
[2] http://unfccc.int/files/paris_agreement/application/pdf/cop_23_outcome-resilience_final.pdf

grains before the peak of the El Niño event, thus reducing crop losses and staving off hunger.[331] Likewise, in 2017 FAO used early warning information to prompt early action in Ethiopia, Kenya and Somalia, lessening the impact of drought on pastoralists by providing thousands of vulnerable families with livestock feed, water and veterinary treatment ahead of the crisis peak.[332]

When integrated into other food security, nutritional or wider poverty-reduction interventions, these systems also represent an opportunity to protect lives and livelihood assets by helping to guarantee food access and stability of food prices. This can include import–export regulations that reduce speculative behaviour through the release of stored food stocks, subsidy programmes for rural incomes, or cash distribution and/or social protection systems, each targeting vulnerable groups at risk of exposure to climate variability and extremes.

Integrating climate risk monitoring into food security and nutrition monitoring is also very important. An example is the multi-stakeholder Integrated Food Security Phase Classification (IPC), which is a set of analytical tools and processes to analyse and classify the severity of acute and chronic food insecurity, specifically designed to provide actionable information to decision-makers in both emergency and development contexts. The IPC analytical framework shown earlier (see Figure 28) has at its core the monitoring and analysis of acute and ongoing events or hazards – including climate variability and extreme climate events – and the analysis of their impact on the food security and nutrition status of the population. Not only does the IPC provide actionable information on current conditions but it also identifies the risk factors to monitor – including seasonal rainfall patterns and the progression of climate events such as droughts – and generates food security projections to inform early warning and action. More than 40 countries worldwide are now implementing the IPC, including countries in Africa, Asia, Central America and the Caribbean, and the Near East.[333]

Emergency preparedness and response

Another important set of tools falls into the category of emergency preparedness and response, which are typical humanitarian actions. Emergency preparedness is an essential element of DRR, helping to reduce the impact of a disaster by building the knowledge and capacities of governments, organizations, communities and individuals to effectively anticipate, respond to and recover from the impacts of disasters (whether they be likely, imminent or current).[334] Measures can include early warning; contingency planning; establishment of multisectoral and sectoral emergency humanitarian coordination mechanisms; exercise management; health service and facility preparedness; allocation of food, seeds and grazing strategic reserves; establishment of safe storage facilities for seeds and harvest; livestock shelters; and safe and hygienic food preparation facilities.[335]

In its work with emergency preparedness and response, WFP integrates climate information into early warning systems, using groundbreaking technology to help forecast emergencies and respond quickly with quality programmes to deliver life-saving food assistance. In 2017, WFP provided in-kind food, vouchers, cash and nutrition support to 9 million people affected by climate-related disasters in the Caribbean, the Horn of Africa and in South Asia. In an effort to deliver on the Core Commitments for Children in the context of climate shocks and other emergency settings, UNICEF has developed specific guidance on the preparedness planning process focusing on children.[336] Emergency preparedness is an important approach because vulnerability reduction measures themselves cannot always avert a crisis.

On the other side of the coin, emergency response to climate-related disasters not only saves lives and livelihoods but is also crucial in ensuring that people do not become irreversibly destitute and dependent on international assistance. Emergency responses should aim to enable people to quickly become self-reliant and resume livelihood activities including local food production and income generation.[337] Disasters can even offer new opportunities to "build back better", whereby people can be assisted to transition from unsustainable practices towards more risk-sensitive and viable management of resources that enhance resilient and sustainable livelihoods.

In the aftermath of Cyclone Pam in Vanuatu in 2015, FAO assisted in the design and building of stronger and safer fishing boats and the introduction of more sustainable and safer fishing practices.[338] Current policies and public and private sector investments in capture fisheries and aquaculture are typically framed around their potential poverty and food insecurity reduction but are rarely viewed through a nutrition-sensitive lens.[339] Supporting households in the aftermath of a climate shock through timely and context-relevant interventions can save livelihoods, which is essential for building climate resilience. For example, households in Kyrgyzstan who were able to restock their herds after a harsh winter in 2012 that killed many livestock were able to increase their food »

BOX 18

HOUSEHOLDS AFFECTED BY CLIMATE SHOCKS WHO ARE ABLE TO RESTOCK OR ACCESS VETERINARY SERVICES HAVE HIGHER FOOD CONSUMPTION IN KYRGYZSTAN

In 2012, a harsh winter in Kyrgyzstan caused many livestock to die, contributing to a significant decline in herding families' food consumption expenditure. Households who were able to replenish their herds after the shock experienced a 10 percent increase in food consumption expenditure in the medium term (four years after the shock) compared to households hit by the shock who were not able to restock (see figure below). Households who had greater access to public veterinary services also reported higher food consumption expenditures compared to households with insufficient access to these services.

This evidence highlights the importance of supporting households in the aftermath of a climate shock through timely and context-relevant interventions to save livelihoods and build resilience. Interventions using cash transfer programmes or ad hoc insurance schemes that facilitate restocking investments could be an option for similar cases. Moreover, facilitating access to veterinary services and vaccines could mitigate the short-term adverse effects of shocks.

Beside immediate livelihood protection interventions, public and private interventions to reduce vulnerabilities in the livestock sector are important for longer-term resilience and sustainability. These measures may include preventing animal losses through the improvement of storage capacities that increase the availability of winter forage in lean winters. Actions should also be combined with efforts to improve the genetic pool of livestock species through breeding programmes that select for resiliency traits. These more resilient animals can be incorporated into local herds and distributed to households living in areas prone to climate shocks and harsh conditions, thereby preparing them for climate variability and extremes.

RESTOCKING AND VETERINARY SERVICES BUILD CLIMATE RESILIENCE AND INCREASE FOOD CONSUMPTION

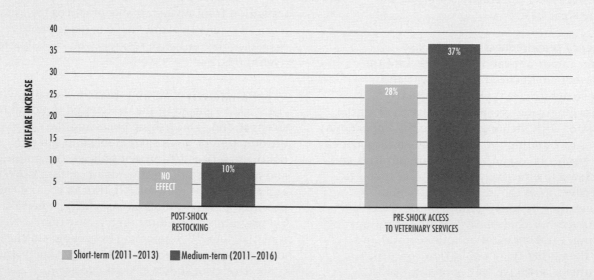

NOTES: Short-term (2011–2013) and medium-term (2011–2016) percent increase in food consumption as a result of restocking and access to veterinary services before and after a harsh winter. All effects are statistically significant, except short-term restocking after the shock of a harsh winter. For this exception it means that it does not have an effect on households' food consumption (no effect). Welfare is defined as per capita household food consumption expenditure.
SOURCE: FAO (forthcoming). *How do extreme weather events affect livestock herders' welfare? Evidence from Kyrgyzstan.* Rome.

» consumption expenditure compared to those who did not (Box 18).

Vulnerability reduction measures

Vulnerability reduction measures combine disaster risk reduction and climate change adaptation practices aiming to prevent and mitigate the impact of climate events and variability at community, farm and agro-ecosystem levels. These include the application of climate-resilient and climate-smart agricultural good practices as well as climate-proofed infrastructures and nature-based solutions and thus need investments at scale. Several examples of vulnerability reduction interventions are given in Boxes 14, 15, 16, 17 and 18.

Shock-responsive social protection, risk transfers and forecast-based financing

Social protection mechanisms can help to reduce disaster risk vulnerability and strengthen livelihoods against the impact of a range of shocks, enabling more people to anticipate risks, bounce back better and faster[340] and become more resilient.

To specifically help individuals and households prepare for and recover from climate variability and extremes, it is necessary to build in the element of "shock-responsiveness" or "adaptiveness" in existing instruments such as cash transfers, pensions and employment guarantee schemes. Importantly, all types of social protection should also be nutrition-sensitive, protect against all forms of malnutrition, explicitly incorporate nutrition objectives and target the nutritionally vulnerable.[341]

Safety nets are a subset of social protection and can be used as direct social assistance instruments for the poor with the aim of responding to and managing climate-related disasters. They include distributing food assistance; subsidizing prices for foodstuffs; providing vouchers, coupons or school meals; and providing support through cash transfers or public works activities. The choice of an instrument or combination of instruments depends on the context and goal.[342] USAID has observed that a package of early humanitarian response and safety nets is about 30 percent more efficient than typical humanitarian aid in Africa,

where a combined resilience-building scenario involving both early humanitarian response and safety nets could save USD 4.3 billion over 15 years.[343]

A joint WFP–Bangladesh Government programme – called Enhancing Resilience (ER) of rural poor communities exposed to climate shocks – has used safety nets to allow participants over the course of two years to build community assets and take part in exchange for cash and food. According to an impact evaluation, participants now are less likely to engage in negative coping strategies than non-participants.[344] In 2017, FAO provided a package in Somalia consisting of cash transfers, quality local seeds, land preparation and irrigation support, training and safe storage equipment. This helped families buy food and meet immediate needs while growing food over the medium to longer term.[345]

Risk transfers can also help significantly reduce (though not fully eliminate) the negative impacts of climate variability and extremes. Vulnerable people including small-scale family farmers are often faced with significant uncertainty, which prompts them to invest in low-risk production assets and technology at the expense of profitability or to allocate labour into less profitable off-farm activities. These risk-adverse activities maintain and can even worsen farmer families' vulnerable conditions with regard to food security and nutrition.

Recent innovative solutions of risk transfer, such as climate risk insurance and forecast-based financing, are helping to formally or informally shift the financial consequences of particular risks from one party to another, at the level of the household, community, enterprise or state.[346]

Climate risk insurance protects people, businesses and states from the adverse effects of climate variability and extremes and reduces the burden of the individual, as risks are borne by defined communities, even before potential damage occurs.[347] An example is the WFP-Oxfam's Rural Resilience Initiative (R4), which since 2016 has provided approximately 37 000 farmers in Ethiopia, Malawi, Senegal and Zambia with index-based insurance climate events. R4 includes international reinsurers, local-level micro-insurance companies and government policies, and implements a climate and weather insurance social protection programme. In Ethiopia, farmers can buy insurance by working additional days in the country's largest public works programme, the Productive Safety Net Programme (PSNP). Between 2015 and 2016, over USD 450 000 in insurance payouts were distributed to small-scale family farmers participating in R4 in Ethiopia, Malawi and Senegal, in response to the dry conditions caused by El Niño.

Forecast-based financing programmes are also available to support preemptive, rapid responses to climate disasters by releasing humanitarian funding, using forecast information for pre-agreed activities, or using Early Action Protocols (EAPs) to define associated roles and responsibilities to reduce risks and enhance preparedness and response.[348] During the 2015–2016 El Niño, for example, WFP's Food Security Climate Resilience Facility (FoodSECuRE) used seasonal climate forecasts to trigger contingency funding for community-level resilience activities before the anticipated shock (drought) occurred, thereby helping preserve food security. In Zimbabwe, WFP and FAO, with the Ministry of Agriculture's extension service (Agritex), field tested the FoodSECuRE Early Action modality in five wards of the Mwenezi district to bolster the resilience capacity of affected small family farmer households by promoting the cultivation of drought-tolerant small grains.

Climate risk and disaster governance
Improving agriculture livelihoods, food security, nutrition and health in the face of climate variability and extremes will only be possible by strengthening governance structures in the environment–food–health nexus. This implies the inclusion of immediate and long-term agriculture, food security, nutrition and health considerations into climate resilience policies, legislation and the larger enabling environment for governance. In this way only will the cross-cutting factors discussed above lead to successful policies and practices addressing climate risks across and within sectors.

Undoubtedly, the fact that existing global policies and strategies are compartmentalized into several dialogues as noted above must be addressed, particularly to support efforts at country and community level. At country level, well-established legislation, institutional structures, policies and plans can create an enabling environment to limit the impact of climate-related disasters and climate variability and build climate resilience. A mix of different tools – including regulation, fiscal instruments, investments in research and knowledge dissemination, support for market accessibility, improvements in infrastructure, and social protection – is seen as being more effective and sustainable in creating a pathway for climate resilience than a single intervention.[349]

Collaboration between the public sector, the private sector and communities is key to ensuring comprehensive, coherent and complementary actions. The Pacific SIDS, which are particularly vulnerable to tropical cyclones, droughts and floods (Box 10), provide a good example of climate risk and disaster governance at the national and regional level within the context of sustainable development (Box 19).

In Vanuatu, for example, there is some integration between sectors on managing climate shocks and climate change, such as national clusters – including the food security cluster, the health cluster and the gender and social protection cluster – that have representatives from different government ministries and NGO and CSO partners. Negotiations are currently underway between the Ministry of Agriculture and the Ministry of Health to develop a Memorandum of Understanding to jointly work on climate issues. However, several important barriers to operationalizing these governance structures exist and need to be addressed.

One of the most significant challenges is limited local capacity. Vanuatu has a small human resource base that is already widely stretched without the added burden of addressing climate risks, both short- and long-term. Long-term strategic climate shock and climate change management planning is difficult in a country overwhelmed by a high

BOX 19
CLIMATE RESILIENCE IN PACIFIC SMALL ISLAND DEVELOPING STATES (SIDS)

The Secretariat of the Pacific Community (SPC) has been helping to support climate change adaptation and disaster risk governance in the Pacific SIDS. SPC is the largest scientific and technical international organization in the Pacific, working at both the regional and national level. It brings together the leadership and guidance of member countries and territories for the design and implementation of multisectoral responses aligned to national priorities, including application of high-quality scientific and technical knowledge and innovation on DRR and CCA at the community and country level.

Their work includes providing assistance for gender mainstreaming activities, developing policies and legislation, and improving the capacity of governments and civil society for advocacy and for monitoring the implementation of human rights standards.[1]

At the regional level, the Framework for Resilient Development in the Pacific (FRDP) 2017–2030 supports the overall objective of strengthening the resilience of Pacific Island Communities to the impacts of slow- and sudden-onset natural hazards. The FRDP identifies three goals linking humanitarian and development interventions:

1. strengthened integrated risk management to enhance climate and disaster resilience;
2. low carbon development; and
3. strengthened disaster preparedness, response and recovery.

In addition, the Pacific Resilience Partnership (PRP) was established to provide the governance structure for FRDP implementation support and monitoring.[2] Through the crisis and disaster governance structures and frameworks, different members are applying high-quality scientific and technical knowledge and innovation on DRR/DRM and CCA at the community and country level to increase the resilience of livelihoods.

Work also includes assisting gender mainstreaming activities, developing policies and legislation, and improving the capacity of governments and civil society for advocacy and for monitoring the implementation of human rights standards. The Pacific region also possesses an intricate network of national institutions and regional initiatives that complement each other. For example, the Online Climate Outlook Forum (OCOF) offers a forum for Pacific Island Meteorological Services; and the Pacific Catastrophe Risk Assessment and Financing Initiative (PCRAFI) facility was established in June 2016 by the Pacific Islands' Ministers of Finance to put the Pacific countries at the forefront of efforts to further expand disaster risk finance.[3] The Pacific Catastrophe Risk Insurance Pilot was introduced to provide parametric catastrophe insurance against tropical cyclones and earthquakes, demonstrating that risk insurance is a viable vulnerability reduction measure for the region. Thanks to risk diversification and economies of scale derived from risk pooling across multiple Pacific SIDS, this initiative has been shown to reduce the cost of reinsurance by up to 50 percent compared to individual purchases of comparable coverage.[4]

SOURCES:

[1] Pacific Community (SPC). 2015. *Pacific Community Strategic Plan 2016–2020: Sustainable Pacific development through science, knowledge and innovation.* Nouméa, France.

[2] Pacific Resilience Partnership (PRP). 2017. *Pacific Resilience Partnership (PRP) Governance Structure - PRP Working Group Draft Final* [online]. www.pacificmet.net/sites/default/files/inline-files/documents/WP%208.0%20Att%202-PRP%20Working%20Group%20Governance%20Paper%20clean%2016%20June.pdf

[3] World Bank. 2017. Pacific Islands Take the Lead on Financial Protection from Disasters. In: *The World Bank* [online]. Washington, DC. www.worldbank.org/en/news/press-release/2017/03/31/pacific-islands-take-the-lead-on-financial-protection-from-disasters

[4] B. Lucas. 2015. *Disaster risk financing and insurance in the Pacific* (GSDRC Helpdesk Research Report 1314). Birmingham, UK, University of Birmingham.

frequency of natural hazards, including regular cyclones and droughts. It is ironic that because of extreme climate events, staff are prevented from dedicating time to long-term strategic planning and response

management to these events. A local climate change adaptation expert noted that, "because we've had so many events, they just get swamped in dealing with one disaster after the other."[350] ∎

2.5 OVERALL CONCLUSION

This second and final part of the report sends a clear message that climate variability and exposure to more complex, frequent and intense climate extremes are threatening to erode and reverse gains made in ending hunger and malnutrition. Evidence shows that, for many countries, recent increases in hunger are associated with extreme climate events, especially where there is both high exposure to climate extremes and high vulnerability related to agriculture and livelihood systems.

Climate variability and extremes — in addition to conflict and violence in some parts of the world — are a key driver behind the recent rises in global hunger identified in Part 1 of the report and one of the leading causes of severe food crises. They are negatively impacting livelihoods and all dimensions of food security (availability, access, utilization and stability), as well as contributing to the other underlying causes of malnutrition related to child care and feeding, health services and environmental health. The risk of food insecurity and malnutrition is greater nowadays because livelihoods and livelihood assets – especially of the poor – are more exposed and vulnerable to climate variability and extremes. What can be done to prevent this threat from eroding and reversing gains made in ending hunger and malnutrition over recent years?

This second part of the report sends an urgent call out to accelerate and scale up actions to strengthen resilience and adaptive capacity to climate variability and extremes. There is a pressing need to increase resilience in a broad sense – i.e. resilience of people's agricultural livelihoods, food systems and nutrition through climate resilience strategies, programmes and investments that address not only the direct impacts but also the underlying vulnerabilities, which in most cases are aggravated by the changing nature of climate variability and extremes.

National and local governments are facing challenges in trying to identify measures to prevent and reduce risk and address the effects of increased climate variability and extremes. They can be guided by existing global policy platforms whereby climate resilience is an important element: climate change (governed by the UNFCCC and the 2015 Paris Agreement), disaster risk reduction (the Sendai Framework on Disaster Risk Reduction), humanitarian emergency response (The 2016 World Humanitarian Summit and the Grand Bargain), improved nutrition and healthy diets (the Second International Conference on Nutrition [ICN2] and the UN Decade of Action on Nutrition 2016–2025) and development as part of the overarching 2030 Agenda for Sustainable Development.

However, it is important to ensure better integration of these global policy platforms to ensure actions across and within sectors such as environment, food, agriculture and health, pursue coherent objectives and actions. Challenges include poorly defined institutional roles between different ministries, capacity gaps, compartmentalized approaches and actions, adaptation and risk management constraints, and a lack of technical capacity and data. These challenges are magnified by the comprehensiveness of livelihoods and food systems and the interrelated nature of climate, food security, nutrition and health issues.

The success of climate resilience policies, programmes and practices requires renewed efforts and new approaches that help people anticipate, absorb and adapt to climate variability and extremes. There are a number of cross-cutting factors that are critical, as well as tools and interventions that are adaptable to specific contexts:

▶ *Cross-cutting factors* that influence the whole livelihoods and food system including climate risk assessments, science and interdisciplinary knowledge, participatory and inclusive approaches, a user-focused approach centred on climate-vulnerable groups, as well as dependable, multi-year and large-scale funding for stepping up climate-resilient investments for agriculture (including crop, livestock, fisheries, aquaculture, and forestry subsectors), food security and nutrition.

▶ *A set of tools and interventions* that makes it possible to implement climate resilience policies, programmes and practices such as risk monitoring and early warning systems; emergency preparedness and response; vulnerability reduction measures, shock-responsive social protection, risk transfers and forecast-based financing; and strengthened governance structures in the environment–food–health nexus.

These policy directions are essential to respond to this report's urgent call to accelerate and scale up actions to strengthen resilience and adaptive capacity to climate variability and extremes. Otherwise the goals of ending hunger and malnutrition in all forms by 2030 (SDG Targets 2.1 and 2.2) along with other goals – such as taking action to combat climate change and its impacts (SDG13) – will remain elusive. ∎

**SOMALI REGION,
ETHIOPIA**
Female farmers in the
drought-affected Somali
Region, Ethiopia, where FAO,
IFAD and WFP are engaged
in complementary projects to
boost productivity, strengthen
livelihoods and improve
nutrition.
©FAO/IFAD/WFP/
Michael Tewelde

ANNEXES

ANNEX 1

STATISTICAL TABLES AND METHODOLOGICAL NOTES TO PART 1

TABLE A1.1
PROGRESS TOWARDS THE SUSTAINABLE DEVELOPMENT GOALS (SDGs): PREVALENCE OF UNDERNOURISHMENT, SEVERE FOOD INSECURITY, SELECTED FORMS OF MALNUTRITION AND EXCLUSIVE BREASTFEEDING

REGIONS/SUBREGIONS/COUNTRIES	PREVALENCE OF UNDERNOURISHMENT IN THE TOTAL POPULATION[1]		PREVALENCE OF SEVERE FOOD INSECURITY IN THE TOTAL POPULATION[1,2]	PREVALENCE OF WASTING IN CHILDREN (UNDER 5 YEARS OF AGE)	PREVALENCE OF STUNTING IN CHILDREN (UNDER 5 YEARS OF AGE)		PREVALENCE OF OVERWEIGHT IN CHILDREN (UNDER 5 YEARS OF AGE)		PREVALENCE OF OBESITY IN THE ADULT POPULATION (18 YEARS AND OLDER)		PREVALENCE OF ANAEMIA AMONG WOMEN OF REPRODUCTIVE AGE (15–49)		PREVALENCE OF EXCLUSIVE BREASTFEEDING AMONG INFANTS 0–5 MONTHS OF AGE	
	2004–06	2015–17	2015–17	2017[3]	2012[4]	2017[3]	2012[4]	2017[3]	2012	2016	2012	2016[5]	2012[6]	2017[7]
	%	%	%	%	%	%	%	%	%	%	%	%	%	%
WORLD	14.3	10.8	9.2	7.5	24.9	22.2	5.4	5.6	11.7	13.2	30.3	32.8	36.9	40.7
Least developed countries	28.6	24.2	23.5	8.5	36.9	33.7	3.4	4.0	4.4	5.4	39.3	39.6	44.8	50.6
Landlocked developing countries	26.4	22.4	20.1	6.9	35.4	31.8	3.9	4.0	7.3	8.3	32.1	33.1	45.5	53.1
Small island developing states	21.1	17.1	n.a.	n.a.	n.a.	n.a.	n.a.	n.a.	18.6	20.9	30.0	31.5	36.4	31.5
Low-income economies	31.6	28.2	n.a.	7.4	38.6	35.2	3.3	3.2	4.7	5.7	37.4	37.3	43.6	51.0
Lower-middle-income economies	19.4	14.1	n.a.	11.5	35.4	31.5	3.7	3.9	6.1	7.3	42.2	43.0	39.4	46.0
Low-income food-deficit countries	22.7	18.5	n.a.	11.7	38.4	34.4	3.1	3.1	12.3	14.2	46.2	46.3	40.8	47.8
AFRICA	21.3	19.6	25.9	7.1	32.6	30.3	5.0	5.0	10.4	11.8	37.7	37.7	35.6	43.5
Northern Africa	6.1	8.4	11.4	8.1	19.1	17.3	9.6	10.3	22.5	25.4	30.9	31.8	40.5	44.4
Algeria	8.8	4.7		n.a.	11.7	n.a.	12.4	n.a.	23.1	26.6	33.6	35.7	25.4	n.a.
Egypt	5.4	4.8	10.1ª	9.5	30.7	22.3	20.5	15.7	27.9	31.1	29.3	28.5	52.8	39.5
Libya	n.a.	n.a.	n.a.	n.a.	21.0	n.a.	22.4	n.a.	28.3	31.8	30.5	32.5	n.a.	n.a.
Morocco	5.7	3.9	n.a.	n.a.	14.9	n.a.	10.7	n.a.	22.4	25.6	34.2	36.9	27.8	n.a.
Sudan	--	25.2	n.a.	16.3	34.1	38.2	1.5	3.0	5.6	7.4	29.4	30.7	41.0	54.6
Tunisia	5.6	4.9	n.a.	n.a.	10.1	n.a.	14.3	n.a.	24.1	27.3	28.1	31.2	8.5	n.a.

TABLE A1.1
((CONTINUED)

REGIONS/SUBREGIONS/COUNTRIES	PREVALENCE OF UNDERNOURISHMENT IN THE TOTAL POPULATION[1] (%)		PREVALENCE OF SEVERE FOOD INSECURITY IN THE TOTAL POPULATION[1,2] (%)	PREVALENCE OF WASTING IN CHILDREN (UNDER 5 YEARS OF AGE) (%)	PREVALENCE OF STUNTING IN CHILDREN (UNDER 5 YEARS OF AGE) (%)		PREVALENCE OF OVERWEIGHT IN CHILDREN (UNDER 5 YEARS OF AGE) (%)		PREVALENCE OF OBESITY IN THE ADULT POPULATION (18 YEARS AND OLDER) (%)		PREVALENCE OF ANAEMIA AMONG WOMEN OF REPRODUCTIVE AGE (15–49) (%)		PREVALENCE OF EXCLUSIVE BREASTFEEDING AMONG INFANTS 0–5 MONTHS OF AGE (%)	
	2004–06	2015–17	2015–17	2017[3]	2012[4]	2017[3]	2012[4]	2017[3]	2012	2016	2012	2016[5]	2012[6]	2017
Northern Africa (excluding Sudan)	6.1	4.8	8.7	n.a.	n.a.	n.a.	n.a.	n.a.	25.4	28.7	31.2	32.1	40.4	39.5
Sub-Saharan Africa	24.4	22.2	29.2	6.9	35.0	32.6	4.2	4.1	6.9	8.0	39.5	39.2	34.8	43.4
Eastern Africa	34.4	31.2	29.2	6.0	38.5	35.6	4.5	4.4	4.3	5.2	30.6	31.2	48.7	59.8
Burundi	n.a.	n.a.		5.1	57.5	55.9	2.9	1.4	3.5	4.4	25.6	26.7	69.3	82.3
Comoros	n.a.	n.a.	n.a.	n.a.	32.1	n.a.	10.9	n.a.	5.8	6.9	27.6	29.3	11.4	n.a.
Djibouti	32.2	19.7	n.a.	n.a.	33.5	n.a.	8.1	n.a.	10.8	12.2	30.9	32.7	12.4	n.a.
Eritrea	n.a.	n.a.	n.a.	n.a.	50.3	n.a.	1.9	n.a.	3.2	4.1	36.9	38.1	68.7	n.a.
Ethiopia	39.7	21.4		9.9	44.2	38.4	1.8	2.8	2.9	3.6	21.7	23.4	52.0	56.5
Kenya	28.2	24.2	35.6	4.0	35.2	26.0	5.0	4.1	4.8	6.0	27.5	27.2	31.9	61.4
Madagascar	35.0	43.1	n.a.	n.a.	49.2	n.a.	n.a.	n.a.	3.6	4.5	36.6	36.8	41.9	n.a.
Malawi	26.1	26.3	52.4	2.7	47.8	37.1	9.2	4.5	3.9	4.7	32.3	34.4	70.8	59.4
Mauritius	5.2	5.8	5.9	n.a.	n.a.	n.a.	n.a.	n.a.	10.1	11.5	21.6	25.1	n.a.	n.a.
Mozambique	37.0	30.5		n.a.	43.1	n.a.	7.9	n.a.	5.1	6.0	49.9	51.0	40.0	41.0
Rwanda	45.3	36.1		1.7	44.3	36.7	7.1	7.7	3.8	4.8	19.4	22.3	83.8	86.9
Seychelles	n.a.	n.a.	2.4	n.a.	7.9	n.a.	10.2	n.a.	12.5	14.6	20.3	22.3	n.a.	n.a.
Somalia	n.a.	n.a.		n.a.	25.3	n.a.	3.0	n.a.	5.9	6.9	43.5	44.4	5.3	n.a.
South Sudan	--	n.a.		n.a.	31.1	n.a.	6.0	n.a.	n.a.	n.a.	32.3	34.0	44.5	n.a.
Uganda	24.1	41.4		3.6	34.2	28.9	5.8	3.7	5.8	7.1	38.6	37.2	62.3	65.5
United Republic of Tanzania	34.4	32.0		4.5	34.8	34.4	5.5	3.6	3.4	4.1	29.6	28.5	48.7	59.0
Zambia	51.1	44.5		6.3	45.8	40.0	8.4	6.2	5.4	6.5	31.2	33.7	59.9	72.0
Zimbabwe	42.2	46.6		3.2	32.3	26.8	5.8	5.6	11.1	12.3	30.1	28.8	31.3	47.1

REGIONS/SUBREGIONS/COUNTRIES	PREVALENCE OF UNDERNOURISHMENT IN THE TOTAL POPULATION[1]		PREVALENCE OF SEVERE FOOD INSECURITY IN THE TOTAL POPULATION[1,2]	PREVALENCE OF WASTING IN CHILDREN (UNDER 5 YEARS OF AGE)	PREVALENCE OF STUNTING IN CHILDREN (UNDER 5 YEARS OF AGE)		PREVALENCE OF OVERWEIGHT IN CHILDREN (UNDER 5 YEARS OF AGE)		PREVALENCE OF OBESITY IN THE ADULT POPULATION (18 YEARS AND OLDER)		PREVALENCE OF ANAEMIA AMONG WOMEN OF REPRODUCTIVE AGE (15–49)		PREVALENCE OF EXCLUSIVE BREASTFEEDING AMONG INFANTS 0–5 MONTHS OF AGE	
	2004–06	2015–17	2015–17	2017[3]	2012[4]	2017[3]	2012[4]	2017[3]	2012	2016	2012	2016[5]	2012[6]	2017[7]
	%	%	%	%	%	%	%	%	%	%	%	%	%	%
Middle Africa	**32.5**	**25.3**	**39.5**	**7.1**	**34.4**	**32.1**	**4.6**	**4.7**	**5.5**	**6.6**	**45.4**	**43.5**	**28.5**	**37.7**
Angola	54.8	23.9		4.9	29.2	37.6	n.a.	3.3	5.6	6.8	47.3	47.7	n.a.	37.4
Cameroon	20.3	7.3	35.5	5.2	32.6	31.7	6.5	6.7	8.1	9.5	41.7	41.4	19.9	28.0
Central African Republic	39.5	61.8	n.a.	n.a.	40.7	n.a.	1.8	n.a.	5.3	6.3	46.2	46.0	33.0	n.a.
Chad	39.2	39.7	n.a.	13.0	38.7	39.9	2.8	2.5	4.0	4.8	48.1	47.7	3.2	0.1
Congo	40.2	37.5	n.a.	8.2	25.0	21.2	3.6	5.9	7.1	8.4	53.8	51.9	20.2	32.9
Democratic Republic of the Congo	n.a.	n.a.	n.a.	8.1	43.5	42.6	4.9	4.4	4.6	5.6	44.7	41.0	36.4	47.3
Equatorial Guinea	n.a.	n.a.	n.a.	n.a.	26.2	n.a.	9.7	n.a.	6.2	7.4	44.1	43.7	7.4	n.a.
Gabon	10.9	9.4	n.a.	n.a.	17.5	n.a.	7.7	n.a.	12.0	13.4	58.3	59.1	5.1	n.a.
Sao Tome and Principe	9.4	10.2	n.a.	4.0	31.6	17.2	11.6	2.4	8.9	10.6	45.4	46.1	50.3	71.7
Southern Africa	**6.5**	**8.1**	**27.3**	**4.0**	**30.2**	**29.1**	**12.6**	**13.7**	**23.2**	**25.6**	**25.9**	**26.0**	**n.a.**	**35.0**
Botswana	31.9	28.5	39.8	n.a.	31.4	n.a.	11.2	n.a.	14.7	16.1	29.4	30.2	20.3	n.a.
Eswatini	17.0	20.7	n.a.	2.0	31.0	25.5	10.7	9.0	12.0	13.5	26.7	27.2	43.8	63.8
Lesotho	11.7	12.8	50.0	2.8	39.0	33.2	7.3	7.4	12.0	13.5	27.2	27.4	52.9	66.9
Namibia	25.1	25.4	39.8	7.1	29.6	23.1	4.6	4.1	12.9	15.0	24.7	23.2	22.1	48.3
South Africa	4.4	6.1	n.a.	2.5	27.2	27.4	17.2	13.3	24.5	27.0	25.7	25.8	n.a.	31.6
Western Africa	**12.3**	**13.1**	**25.1**	**8.1**	**31.9**	**29.9**	**2.6**	**2.4**	**6.4**	**7.7**	**50.0**	**49.3**	**22.1**	**31.0**
Benin	15.4	10.4	n.a.	4.5	44.7	34.0	11.4	1.7	7.0	8.2	51.5	46.9	32.5	41.4
Burkina Faso	24.9	21.3	23.8	7.6	32.9	27.3	2.8	1.2	3.6	4.5	50.5	49.6	38.2	50.1
Cabo Verde	14.0	12.3	n.a.	n.a.	n.a.	n.a.	n.a.	n.a.	8.9	10.6	31.2	33.3	59.6	n.a.
Côte d'Ivoire	20.0	20.7	n.a.	6.0	29.6	21.6	3.2	1.5	7.6	9.0	51.8	52.9	11.8	23.5
Gambia	15.1	9.6	25.9	11.1	21.2	25.0	1.1	3.2	7.3	8.7	57.2	57.5	33.5	46.8
Ghana	9.3	6.1	7.9	4.7	22.7	18.8	2.6	2.6	8.3	9.7	48.6	46.4	45.7	52.1

TABLE A1.1 (CONTINUED)

REGIONS/SUBREGIONS/COUNTRIES	PREVALENCE OF UNDERNOURISHMENT IN THE TOTAL POPULATION[1] (%) 2004-06	2015-17	PREVALENCE OF SEVERE FOOD INSECURITY IN THE TOTAL POPULATION[1,2] (%) 2015-17	PREVALENCE OF WASTING IN CHILDREN (UNDER 5 YEARS OF AGE) (%) 2017[3]	PREVALENCE OF STUNTING IN CHILDREN (UNDER 5 YEARS OF AGE) (%) 2012[4]	2017[3]	PREVALENCE OF OVERWEIGHT IN CHILDREN (UNDER 5 YEARS OF AGE) (%) 2012[4]	2017[3]	PREVALENCE OF OBESITY IN THE ADULT POPULATION (18 YEARS AND OLDER) (%) 2012	2016	PREVALENCE OF ANAEMIA AMONG WOMEN OF REPRODUCTIVE AGE (15-49) (%) 2012	2016[5]	PREVALENCE OF EXCLUSIVE BREASTFEEDING AMONG INFANTS 0-5 MONTHS OF AGE (%) 2012[6]	2017[7]
Guinea	21.3	19.7	39.9	8.1	31.3	32.4	3.8	4.0	5.5	6.6	50.9	50.6	20.4	35.2
Guinea-Bissau	24.4	26.0	n.a.	6.0	32.2	27.6	3.2	2.3	6.8	8.2	44.0	43.8	38.3	52.5
Liberia	39.4	38.8	56.6	5.6	41.8	32.1	4.2	3.2	7.3	8.6	37.3	34.7	27.8	54.6
Mali	11.1	6.0	n.a.	13.5	27.8	30.4	1.0	1.9	5.9	7.1	54.8	51.3	20.2	37.3
Mauritania	12.1	11.3	37.2	14.8	22.0	27.9	1.2	1.3	9.7	11.3	37.2	37.2	26.7	41.4
Niger	15.1	14.4	24.8	10.3	43.0	42.2	3.0	n.a.	3.9	4.7	49.2	49.5	23.3	n.a.
Nigeria	6.5	11.5		10.8	36.0	43.6	3.0	1.5	6.4	7.8	49.9	49.8	14.7	23.3
Senegal	21.6	11.3		7.2	15.5	17.0	0.7	0.9	6.2	7.4	53.5	49.9	37.5	36.4
Sierra Leone	37.0	25.5		9.4	44.9	37.9	10.3	8.9	6.3	7.5	47.9	48.0	31.2	31.4
Togo	26.0	16.2	30.5	6.7	29.8	27.5	1.6	2.0	5.9	7.1	50.0	48.9	62.1	57.2
Sub-Saharan Africa (including Sudan)	**24.4**	**22.3**	**29.0**	**n.a.**	**n.a.**	**n.a.**	**n.a.**	**n.a.**	**6.8**	**7.9**	**39.1**	**38.8**	**35.0**	**43.8**
ASIA*	**17.1**	**11.5**	**6.7**	**9.7**	**27.1**	**23.2**	**4.5**	**4.8**	**6.0**	**7.3**	**33.5**	**36.6**	**38.8**	**40.1**
Central Asia	**11.0**	**6.0**	**2.6**	**3.7[b]**	**15.5**	**11.8[b]**	**10.1**	**10.7[b]**	**14.4**	**16.8**	**33.2**	**33.8**	**29.2**	**41.0**
Kazakhstan	5.9	<2.5	1.4	3.1	13.1	8.0	13.3	9.3	18.7	21.3	29.4	30.7	31.8	37.8
Kyrgyzstan	9.7	6.5		2.8	17.8	12.9	9.0	7.0	12.9	15.4	32.1	36.2	56.0	40.9
Tajikistan	n.a.	n.a.	7.8	n.a.	26.8	n.a.	6.6	n.a.	10.4	12.6	29.7	30.5	32.6	35.8
Turkmenistan	4.8	5.5	n.a.	4.2	18.9	11.5	4.5	5.9	14.9	17.5	31.1	32.6	10.9	58.3
Uzbekistan	14.5	7.4		n.a.	19.6	n.a.	12.8	n.a.	12.9	15.3	36.8	36.2	23.8	n.a.
Eastern Asia*	**14.0**	**8.5**	**0.8**	**1.8**	**7.9**	**5.3**	**5.5**	**5.2**	**5.0**	**6.4**	**20.8**	**26.1**	**28.6**	**18.7**
China	15.2	8.7		1.9	9.4	8.1	6.6	n.a.	5.1	6.6	20.7	26.4	27.6	18.6
China, mainland	*15.5*	*8.8*		*n.a.*	*n.a.*	*n.a.*	*n.a.*	*n.a.*	*n.a.*	*n.a.*	*n.a.*	*n.a.*	*n.a.*	*n.a.*
Taiwan Province of China	*4.7*	*3.4*		*n.a.*	*n.a.*	*n.a.*	*n.a.*	*n.a.*	*n.a.*	*n.a.*	*n.a.*	*n.a.*	*n.a.*	*n.a.*

REGIONS/SUBREGIONS/COUNTRIES	PREVALENCE OF UNDERNOURISHMENT IN THE TOTAL POPULATION[1] (%)		PREVALENCE OF SEVERE FOOD INSECURITY IN THE TOTAL POPULATION[1,2] (%)	PREVALENCE OF WASTING IN CHILDREN (UNDER 5 YEARS OF AGE) (%)	PREVALENCE OF STUNTING IN CHILDREN (UNDER 5 YEARS OF AGE) (%)		PREVALENCE OF OVERWEIGHT IN CHILDREN (UNDER 5 YEARS OF AGE) (%)		PREVALENCE OF OBESITY IN THE ADULT POPULATION (18 YEARS AND OLDER) (%)		PREVALENCE OF ANAEMIA AMONG WOMEN OF REPRODUCTIVE AGE (15-49) (%)		PREVALENCE OF EXCLUSIVE BREASTFEEDING AMONG INFANTS 0-5 MONTHS OF AGE (%)	
	2004–06	2015–17	2015–17	2017[3]	2012[4]	2017[3]	2012[4]	2017[3]	2012	2016	2012	2016[5]	2012[6]	2017[7]
China, Hong Kong SAR	<2.5	<2.5	n.a.	n.a.	n.a.	n.a.	n.a.	n.a.	n.a.	n.a.	n.a.	n.a.	n.a.	n.a.
China, Macao SAR	14.6	11.4	n.a.	n.a.	n.a.	n.a.	n.a.	n.a.	n.a.	n.a.	n.a.	n.a.	n.a.	n.a.
Democratic People's Republic of Korea	35.4	43.4	n.a.	n.a.	27.9	n.a.	0.0	n.a.	6.1	7.1	30.0	32.5	68.9	n.a.
Japan	<2.5	<2.5	<0.5	n.a.	7.1	n.a.	1.5	n.a.	3.8	4.4	19.4	21.5	n.a.	n.a.
Mongolia	31.0	18.7	2.8	1.0	15.6	10.8	6.7	10.5	16.3	19.6	16.3	19.5	65.7	46.0
Republic of Korea	<2.5	<2.5	<0.5[c]	n.a.	2.5	n.a.	7.3	n.a.	4.4	4.9	18.4	22.7	n.a.	n.a.
Eastern Asia (excluding China, mainland)	**5.8**	**6.9**	**<0.5**	**n.a.**	**n.a.**	**n.a.**	**n.a.**	**n.a.**	**n.a.**	**n.a.**	**n.a.**	**n.a.**	**n.a.**	**n.a.**
South-eastern Asia	**18.0**	**9.7**	**8.7**	**8.7**	**29.1**	**25.7**	**5.7**	**7.3**	**5.3**	**6.7**	**25.9**	**28.3**	**33.5**	**n.a.**
Brunei Darussalam	<2.5	2.6	n.a.	n.a.	19.7	n.a.	8.3	n.a.	12.3	14.7	13.9	16.9	n.a.	n.a.
Cambodia	20.0	18.5	14.4	9.6	40.9	32.4	1.9	2.0	2.7	3.5	46.0	46.8	72.8	65.2
Indonesia	18.5	7.7	n.a.	13.5	39.2	36.4	12.3	11.5	5.4	6.9	26.2	28.8	40.9	n.a.
Lao People's Democratic Republic	27.0	16.6	n.a.	n.a.	43.8	n.a.	2.0	n.a.	3.4	4.5	36.5	39.7	39.7	n.a.
Malaysia	3.9	2.9	n.a.	11.5	17.2	20.7	n.a.	6.0	12.7	15.3	22.2	24.9	n.a.	n.a.
Myanmar	32.0	10.5	n.a.	7.0	35.1	29.2	2.6	1.3	4.4	5.7	41.7	46.3	23.6	51.2
Philippines	16.3	13.7	12.9	7.1	33.6	33.4	4.3	3.9	5.0	6.0	18.0	15.7	33.0	n.a.
Singapore	n.a.	n.a.	0.6	n.a.	n.a.	n.a.	n.a.	n.a.	6.1	6.6	19.0	22.2	n.a.	n.a.
Thailand	12.5	9.0	n.a.	5.4	16.3	10.5	10.9	8.2	8.4	10.8	26.3	31.8	12.3	23.1
Timor-Leste	31.3	27.2	n.a.	11.0	57.7	50.2	5.8	1.5	2.4	2.9	33.1	41.3	50.8	50.2
Viet Nam	18.2	10.8	2.3	6.4	23.3	24.6	4.6	5.3	1.5	2.1	21.0	24.2	17.0	24.0
Southern Asia	**21.1**	**15.2**	**10.9**	**15.3**	**37.9**	**33.3**	**2.9**	**3.1**	**4.2**	**5.2**	**48.2**	**48.7**	**46.8**	**52.4**
Afghanistan	33.2	30.3	16.0	9.5	n.a.	40.9	n.a.	5.4	3.7	4.5	37.4	42.0	n.a.	43.1
Bangladesh	16.6	15.2	n.a.	14.3	42.0	36.1	1.6	1.4	2.6	3.4	40.3	39.9	55.9	55.3
Bhutan	n.a.	n.a.	n.a.	n.a.	33.6	n.a.	7.6	n.a.	4.5	5.8	39.2	35.6	48.7	51.4
India	22.2	14.8	n.a.	21.0	47.9	38.4	1.9	2.1	3.0	3.8	51.3	51.4	46.4	54.9

TABLE A1.1 (CONTINUED)

REGIONS/SUBREGIONS/COUNTRIES	PREVALENCE OF UNDERNOURISHMENT IN THE TOTAL POPULATION[1] (%)		PREVALENCE OF SEVERE FOOD INSECURITY IN THE TOTAL POPULATION[1,2] (%)	PREVALENCE OF WASTING IN CHILDREN (UNDER 5 YEARS OF AGE) (%)	PREVALENCE OF STUNTING IN CHILDREN (UNDER 5 YEARS OF AGE) (%)		PREVALENCE OF OVERWEIGHT IN CHILDREN (UNDER 5 YEARS OF AGE) (%)		PREVALENCE OF OBESITY IN THE ADULT POPULATION (18 YEARS AND OLDER) (%)		PREVALENCE OF ANAEMIA AMONG WOMEN OF REPRODUCTIVE AGE (15-49) (%)		PREVALENCE OF EXCLUSIVE BREASTFEEDING AMONG INFANTS 0-5 MONTHS OF AGE (%)	
	2004-06	2015-17	2015-17	2017[3]	2012[4]	2017[3]	2012[4]	2017[3]	2012	2016	2012	2016[5]	2012[6]	2017[7]
Iran (Islamic Republic of)	6.1	4.9		n.a.	6.8	n.a.	n.a.	n.a.	22.0	25.5	27.9	30.5	53.1	n.a.
Maldives	18.2	11.0	n.a.	n.a.	20.3	n.a.	6.5	n.a.	5.9	7.9	41.1	42.6	45.3	n.a.
Nepal	16.0	9.5	7.8	9.7	40.5	35.8	1.5	1.2	3.0	3.8	35.4	35.1	69.6	65.2
Pakistan	23.3	20.5		n.a.	45.0	n.a.	4.8	n.a.	6.3	7.8	50.1	52.1	37.0	37.7
Sri Lanka	18.2	10.9		15.1	14.7	17.3	0.6	2.0	4.3	5.4	30.3	32.6	75.8	82.0
Southern Asia (excluding India)	18.3	16.1	8.6	n.a.	n.a.	n.a.	n.a.	n.a.	n.a.	n.a.	n.a.	n.a.	47.7	46.5
Western Asia	9.5	11.1	9.6	3.9	17.3	15.2	7.7	8.2	25.7	28.6	33.9	36.1	n.a.	n.a.
Armenia	7.8	4.3	3.8	4.2	20.8	9.4	16.8	13.6	18.5	20.9	24.7	29.4	34.1	44.5
Azerbaijan	5.5	<2.5		3.1	16.4	18.0	10.4	13.0	17.1	19.9	36.2	38.5	10.8	12.1
Bahrain	n.a.	n.a.	n.a.	n.a.	n.a.	n.a.	n.a.	n.a.	26.2	28.7	41.4	42.0	n.a.	n.a.
Cyprus	5.7	4.6		n.a.	n.a.	n.a.	n.a.	n.a.	20.9	22.6	21.6	25.2	n.a.	n.a.
Georgia	7.2	7.4	8.9	n.a.	11.3	n.a.	19.9	n.a.	20.3	23.3	25.8	27.5	54.8	n.a.
Iraq	28.2	27.7		n.a.	22.6	n.a.	11.8	n.a.	25.0	27.4	29.0	29.1	19.4	n.a.
Israel	<2.5	<2.5		n.a.	n.a.	n.a.	n.a.	n.a.	25.3	26.7	13.1	15.7	n.a.	n.a.
Jordan	6.6	13.5	13.9	n.a.	7.8	n.a.	4.7	n.a.	30.3	33.4	30.8	34.7	22.7	n.a.
Kuwait	<2.5	<2.5		3.1	4.3	4.9	9.5	6.0	34.3	37.0	21.0	23.8	n.a.	n.a.
Lebanon	3.4	10.9		n.a.	n.a.	n.a.	n.a.	n.a.	28.8	31.3	28.1	31.2	n.a.	n.a.
Oman	10.5	5.4	n.a.	7.5	9.8	14.1	1.7	4.4	20.2	22.9	36.3	38.2	n.a.	32.8
Palestine	n.a.	n.a.	9.2	n.a.	n.a.	n.a.	n.a.	n.a.	n.a.	n.a.	n.a.	n.a.	n.a.	n.a.
Qatar	n.a.	n.a.		n.a.	n.a.	n.a.	n.a.	n.a.	30.6	33.9	25.8	27.7	29.3	n.a.
Saudi Arabia	7.9	5.5		n.a.	9.3	n.a.	6.1	n.a.	31.6	35.0	41.5	42.9	n.a.	n.a.
Syrian Arab Republic	n.a.	n.a.	n.a.	n.a.	27.5	n.a.	17.9	n.a.	22.7	25.8	31.7	33.6	42.6	n.a.

TABLE A1.1 (CONTINUED)

REGIONS/SUBREGIONS/COUNTRIES	PREVALENCE OF UNDERNOURISHMENT IN THE TOTAL POPULATION[1]		PREVALENCE OF SEVERE FOOD INSECURITY IN THE TOTAL POPULATION[1,2]	PREVALENCE OF WASTING IN CHILDREN (UNDER 5 YEARS OF AGE)	PREVALENCE OF STUNTING IN CHILDREN (UNDER 5 YEARS OF AGE)		PREVALENCE OF OVERWEIGHT IN CHILDREN (UNDER 5 YEARS OF AGE)		PREVALENCE OF OBESITY IN THE ADULT POPULATION (18 YEARS AND OLDER)		PREVALENCE OF ANAEMIA AMONG WOMEN OF REPRODUCTIVE AGE (15–49)		PREVALENCE OF EXCLUSIVE BREASTFEEDING AMONG INFANTS 0–5 MONTHS OF AGE	
	2004–06	2015–17	2015–17	2017[3]	2012[4]	2017[3]	2012[4]	2017[3]	2012	2016	2012	2016[5]	2012[6]	2017[7]
	%	%	%	%	%	%	%	%	%	%	%	%	%	%
Turkey	<2.5	<2.5	n.a.	1.7	12.3	9.5	n.a.	10.9	29.0	32.2	29.0	30.9	n.a.	30.1
United Arab Emirates	4.1	2.5	n.a.	n.a.	n.a.	n.a.	1.5	n.a.	24.5	29.9	25.7	27.8	n.a.	n.a.
Yemen	30.1	34.4	n.a.	16.3	46.6	46.5	1.5	2.0	11.8	14.1	65.5	69.6	n.a.	9.7
Central Asia and Southern Asia	20.7	14.8	10.6	14.8	37.0	32.4	3.2	3.4	4.6	5.7	47.7	48.2	46.0	52.1
Eastern Asia and South-eastern Asia*	15.1	8.8	3.0	4.5	16.1	13.2	5.6	6.0	5.1	6.5	22.2	26.7	30.5	21.6
Western Asia and Northern Africa	8.0	9.8	10.4	6.0	18.2	16.3	8.7	9.2	24.2	27.2	32.5	34.1	36.3	35.8
LATIN AMERICA AND THE CARIBBEAN	9.1	6.1	n.a.	1.3	11.4	9.6	7.1	7.3	21.7	24.1	21.2	22.0	30.7	n.a.
Caribbean	23.4	17.2	n.a.	3.2[b]	9.6	8.0[b]	6.6	7.2[b]	21.9	24.8	30.4	31.3	29.0	25.7
Antigua and Barbuda	n.a.	n.a.	n.a.	n.a.	n.a.	n.a.	n.a.	n.a.	17.0	19.1	21.5	22.1	n.a.	n.a.
Bahamas	n.a.	n.a.	n.a.	n.a.	n.a.	n.a.	n.a.	n.a.	29.7	32.1	22.3	23.1	n.a.	n.a.
Barbados	5.9	3.7	n.a.	n.a.	7.7	n.a.	12.2	n.a.	22.2	24.8	20.7	21.6	19.7	n.a.
Cuba	<2.5	<2.5	n.a.	n.a.	n.a.	n.a.	n.a.	n.a.	24.3	26.7	24.3	25.1	48.6	32.8
Dominica	5.7	5.2	n.a.	n.a.	n.a.	n.a.	n.a.	n.a.	25.6	28.2	23.5	24.4	n.a.	n.a.
Dominican Republic	24.4	10.4	n.a.	2.4	10.1	7.1	8.3	7.6	23.5	26.9	29.5	29.7	8.0	4.6
Grenada	n.a.	n.a.	n.a.	n.a.	n.a.	n.a.	n.a.	n.a.	17.5	20.2	22.8	23.5	n.a.	n.a.
Haiti	57.1	45.8	n.a.	n.a.	21.9	n.a.	3.6	n.a.	17.2	20.5	46.1	46.2	39.3	39.9
Jamaica	7.0	8.9	n.a.	3.6	5.7	6.2	7.8	8.5	21.9	24.4	21.8	22.5	23.8	n.a.
Puerto Rico	n.a.		n.a.	n.a.	n.a.	n.a.	n.a.	n.a.	n.a.	23.1	n.a.	n.a.	n.a.	n.a.
Saint Kitts and Nevis	n.a.	n.a.	n.a.	n.a.	n.a.	n.a.	n.a.	n.a.	20.4	23.1	n.a.	n.a.	n.a.	n.a.
Saint Lucia	n.a.	n.a.	4.5	n.a.	2.5	n.a.	6.3	n.a.	17.4	19.8	21.4	21.9	3.5	n.a.
Saint Vincent and the Grenadines	9.1	5.7	n.a.	n.a.	n.a.	n.a.	n.a.	n.a.	20.8	23.8	23.9	24.8	n.a.	n.a.
Trinidad and Tobago	11.8	4.9	n.a.	n.a.	11.0	n.a.	11.5	n.a.	16.7	19.7	21.8	22.5	11.7	n.a.

REGIONS/SUBREGIONS/COUNTRIES	PREVALENCE OF UNDERNOURISHMENT IN THE TOTAL POPULATION[1] (%)		PREVALENCE OF SEVERE FOOD INSECURITY IN THE TOTAL POPULATION[1,2] (%)	PREVALENCE OF WASTING IN CHILDREN (UNDER 5 YEARS OF AGE) (%)	PREVALENCE OF STUNTING IN CHILDREN (UNDER 5 YEARS OF AGE) (%)		PREVALENCE OF OVERWEIGHT IN CHILDREN (UNDER 5 YEARS OF AGE) (%)		PREVALENCE OF OBESITY IN THE ADULT POPULATION (18 YEARS AND OLDER) (%)		PREVALENCE OF ANAEMIA AMONG WOMEN OF REPRODUCTIVE AGE (15–49) (%)		PREVALENCE OF EXCLUSIVE BREASTFEEDING AMONG INFANTS 0–5 MONTHS OF AGE (%)	
	2004–06	2015–17	2015–17	2017[3]	2012[4]	2017[3]	2012[4]	2017[3]	2012	2016	2012	2016[5]	2012[6]	2017[7]
Central America	**8.3**	**6.3**	**10.3**	**0.9**	**16.6**	**14.1**	**6.2**	**6.4**	**24.2**	**26.6**	**15.3**	**15.5**	**21.3**	**33.9**
Belize	4.6	6.5		1.8	19.3	15.0	7.9	7.3	19.9	22.4	21.0	21.7	14.7	33.2
Costa Rica	5.4	4.4	4.8	n.a.	5.6	n.a.	8.1	n.a.	22.4	25.7	13.3	14.9	32.5	n.a.
El Salvador	10.5	10.3	11.7	2.1	20.6	13.6	5.7	6.4	20.4	22.7	18.9	22.7	31.4	46.7
Guatemala	15.8	15.8		0.7	48.0	46.5	4.9	4.7	16.6	18.8	17.5	16.4	49.6	53.2
Honduras	17.0	15.3		n.a.	22.7	n.a.	5.2	n.a.	16.9	19.4	16.3	17.8	30.7	n.a.
Mexico	5.5	3.8	8.9	1.0	13.6	12.4	9.0	5.2	26.0	28.4	14.7	14.6	14.4	30.1
Nicaragua	24.4	16.2		n.a.	17.3	n.a.	8.3	n.a.	19.3	21.8	13.9	16.3	31.7	n.a.
Panama	22.9	9.2		n.a.	19.1	n.a.	n.a.	n.a.	20.2	22.5	24.0	23.4	n.a.	21.5
South America	**7.9**	**4.9**	**6.9**	**1.3[b]**	**9.0**	**7.5[b]**	**7.6**	**7.7[b]**	**20.8**	**23.0**	**22.7**	**23.9**	**36.3**	**n.a.**
Argentina	4.7	3.8	8.7	n.a.	8.2	n.a.	9.9	n.a.	26.3	28.5	15.9	18.6	32.0	n.a.
Bolivia (Plurinational State of)	30.3	19.8		2.0	18.1	16.1	8.7	10.1	16.8	18.7	30.1	30.2	64.3	58.3
Brazil	4.6	<2.5		n.a.	7.1	n.a.	7.3	n.a.	19.9	22.3	25.3	27.2	38.6	n.a.
Chile	3.9	3.3	4.4	0.3	2.0	1.8	9.5	9.3	26.6	28.8	11.6	15.0	n.a.	n.a.
Colombia	9.7	6.5		n.a.	12.7	n.a.	4.8	n.a.	19.9	22.1	22.3	21.1	n.a.	n.a.
Ecuador	17.0	7.8	7.1[c]	1.6	25.2	23.9	7.5	8.0	17.3	19.3	18.4	18.8	n.a.	n.a.
Guyana	9.4	7.5	n.a.	6.4	19.5	12.0	6.7	5.3	16.6	19.2	33.4	32.3	31.3	21.1
Paraguay	11.9	11.2		1.0	10.9	5.6	11.7	12.4	16.7	19.0	20.5	22.8	24.4	29.6
Peru	19.6	8.8		1.0	18.4	13.1	7.2	n.a.	17.2	19.1	20.0	18.5	67.4	69.8
Suriname	10.9	7.6	n.a.	n.a.	8.8	n.a.	4.0	n.a.	24.2	26.5	23.4	24.1	2.8	n.a.
Uruguay	4.3	<2.5		n.a.	10.7	n.a.	7.2	n.a.	26.8	28.9	18.3	20.8	n.a.	n.a.
Venezuela (Bolivarian Republic of)	10.5	11.7		n.a.	13.4	n.a.	6.4	n.a.	23.3	25.2	22.9	23.9	n.a.	n.a.

REGIONS/SUBREGIONS/COUNTRIES	PREVALENCE OF UNDERNOURISHMENT IN THE TOTAL POPULATION[1] (%)		PREVALENCE OF SEVERE FOOD INSECURITY IN THE TOTAL POPULATION[1,2] (%)	PREVALENCE OF WASTING IN CHILDREN (UNDER 5 YEARS OF AGE) (%)	PREVALENCE OF STUNTING IN CHILDREN (UNDER 5 YEARS OF AGE) (%)		PREVALENCE OF OVERWEIGHT IN CHILDREN (UNDER 5 YEARS OF AGE) (%)		PREVALENCE OF OBESITY IN THE ADULT POPULATION (18 YEARS AND OLDER) (%)		PREVALENCE OF ANAEMIA AMONG WOMEN OF REPRODUCTIVE AGE (15-49) (%)		PREVALENCE OF EXCLUSIVE BREASTFEEDING AMONG INFANTS 0–5 MONTHS OF AGE (%)	
	2004–06	2015–17	2015–17	2017[3]	2012[4]	2017[3]	2012[4]	2017[3]	2012	2016	2012	2016[5]	2012[6]	2017[7]
OCEANIA	5.5	6.6	n.a.	n.a.	n.a.	n.a.	n.a.	n.a.	26.5	28.9	14.8	16.5	n.a.	n.a.
Australia and New Zealand	<2.5	<2.5	3.2	n.a.	n.a.	n.a.	n.a.	n.a.	28.2	30.7	8.3	9.5	n.a.	n.a.
Australia	<2.5	<2.5	3.2	n.a.	2.0	n.a.	7.7	n.a.	27.9	30.4	8.1	9.1	n.a.	n.a.
New Zealand	<2.5	<2.5	3.2	n.a.	n.a.	n.a.	n.a.	n.a.	29.5	32.0	9.7	11.6	n.a.	n.a.
Oceania excluding Australia and New Zealand	n.a.	n.a.	n.a.	9.2	37.7	38.1	7.3	8.7	20.1	22.4	33.2	35.4	56.8	n.a.
Melanesia	n.a.	n.a.	n.a.	n.a.	n.a.	n.a.	n.a.	n.a.	18.4	20.7	33.9	35.9	56.8	n.a.
Fiji	4.3	4.4	n.a.	n.a.	n.a.	n.a.	n.a.	n.a.	27.2	30.0	29.8	31.0	n.a.	n.a.
New Caledonia	8.2	11.6	n.a.	n.a.	n.a.	n.a.	n.a.	n.a.	n.a.	n.a.	n.a.	n.a.	n.a.	n.c.
Papua New Guinea	n.a.	n.a.	n.a.	n.a.	49.5	n.a.	13.8	n.a.	17.1	19.4	34.4	36.6	56.1	n.a.
Solomon Islands	11.9	12.3	n.a.	7.9	32.8	31.6	2.5	3.9	17.9	20.5	38.4	38.9	73.7	76.2
Vanuatu	7.0	7.1	n.a.	4.4	25.9	28.5	4.7	4.6	20.7	23.5	24.1	24.0	39.5	72.6
Micronesia	n.a.	n.a.	n.a.	n.a.	n.a.	n.a.	n.a.	n.a.	44.2	46.8	22.3	25.1	69.0	n.a.
Kiribati	4.6	3.1	n.a.	n.a.	n.a.	n.a.	n.a.	n.a.	43.0	45.6	23.8	26.1	69.0	n.a.
Marshall Islands	n.a.	n.a.	n.a.	n.a.	n.a.	n.a.	n.a.	n.a.	50.1	52.4	24.1	26.6	27.3	n.a.
Micronesia (Federated States of)	n.a.	n.a.	n.a.	n.a.	n.a.	n.a.	n.a.	n.a.	38.6	41.6	19.5	23.3	n.a.	n.a.
Nauru	n.a.	n.a.	n.a.	n.a.	24.0	n.a.	2.8	n.a.	59.3	60.7	n.a.	n.a.	67.2	n.a.
Palau	n.a.	n.a.	n.a.	n.a.	n.a.	n.a.	n.a.	n.a.	52.5	54.9	n.a.	n.a.	n.a.	n.a.
Polynesia	3.7	3.9	n.a.	n.a.	n.a.	n.a.	n.a.	n.a.	43.9	46.5	23.0	27.6	51.6	70.3
American Samoa	n.a.	n.a.	n.a.	n.a.	n.a.	n.a.	n.a.	n.a.	n.a.	n.a.	n.a.	n.a.	n.a.	n.a.
Cook Islands	n.a.	n.a.	n.a.	n.a.	n.a.	n.a.	n.a.	n.a.	53.0	55.3	n.a.	n.a.	n.a.	n.a.
French Polynesia	3.9	4.4	n.a.	n.a.	n.a.	n.a.	n.a.	n.a.	n.a.	n.a.	n.a.	n.a.	n.a.	n.a.
Niue	n.a.	n.a.	n.a.	n.a.	n.a.	n.a.	n.a.	n.a.	46.0	49.3	n.a.	n.a.	n.a.	n.a.

TABLE A1.1 (CONTINUED)

REGIONS/SUBREGIONS/COUNTRIES	PREVALENCE OF UNDERNOURISHMENT IN THE TOTAL POPULATION[1] (%)		PREVALENCE OF SEVERE FOOD INSECURITY IN THE TOTAL POPULATION[1,2] (%)	PREVALENCE OF WASTING IN CHILDREN (UNDER 5 YEARS OF AGE) (%)	PREVALENCE OF STUNTING IN CHILDREN (UNDER 5 YEARS OF AGE) (%)		PREVALENCE OF OVERWEIGHT IN CHILDREN (UNDER 5 YEARS OF AGE) (%)		PREVALENCE OF OBESITY IN THE ADULT POPULATION (18 YEARS AND OLDER) (%)		PREVALENCE OF ANAEMIA AMONG WOMEN OF REPRODUCTIVE AGE (15-49) (%)		PREVALENCE OF EXCLUSIVE BREASTFEEDING AMONG INFANTS 0-5 MONTHS OF AGE (%)	
	2004-06	2015-17	2015-17	2017[3]	2012[4]	2017[3]	2012[4]	2017[3]	2012	2016	2012	2016[5]	2012[6]	2017
Samoa	3.5	3.1	n.a.	3.7	n.a.	4.7	n.a.	5.4	42.9	45.5	25.4	31.3	51.3	70.3
Tokelau (Associate Member)	n.a.	n.a.	n.a.	n.a.	n.a.	n.a.	n.a.	n.a.	n.a.	n.a.	n.a.	n.a.	n.a.	n.a.
Tonga	n.a.	n.a.	n.a.	n.a.	8.1	n.a.	17.3	n.a.	43.3	45.9	19.0	21.3	52.2	n.a.
Tuvalu	n.a.	n.a.	n.a.	n.a.	10.0	n.a.	6.3	n.a.	47.8	51.0	n.a.	n.a.	34.7	n.a.
NORTHERN AMERICA AND EUROPE	**<2.5**	**<2.5**	**1.4**	**n.a.**	**n.a.**	**n.a.**	**n.a.**	**n.a.**	**26.7**	**29.0**	**15.4**	**17.8**	**n.a.**	**n.a.**
Northern America	**<2.5**	**<2.5**	**1.1**	**n.a.**	**n.a.**	**n.a.**	**n.a.**	**n.a.**	**34.1**	**36.7**	**10.6**	**12.9**	**25.5**	**26.4**
Bermuda	n.a.	n.a.	n.a.	n.a.	n.a.	n.a.	n.a.	n.a.	n.a.	n.a.	n.a.	n.a.	n.a.	n.a.
Canada	<2.5	<2.5	n.a.	n.a.	n.a.	n.a.	n.a.	n.a.	28.8	31.3	8.5	9.5	n.a.	n.a.
Greenland	n.a.	n.a.	n.a.	n.a.	n.a.	n.a.	n.a.	n.a.	n.a.	n.a.	n.a.	n.a.	n.a.	n.a.
United States of America	<2.5	<2.5	1.1	n.a.	2.1	n.a.	6.0	n.a.	34.7	37.3	10.9	13.3	25.5	26.4
Europe	**<2.5**	**<2.5**	**1.5**	**n.a.**	**n.a.**	**n.a.**	**n.a.**	**n.a.**	**23.4**	**25.4**	**17.6**	**20.2**	**n.a.**	**n.a.**
Eastern Europe	**<2.5**	**<2.5**	**1.2**	**n.a.**	**n.a.**	**n.a.**	**n.a.**	**n.a.**	**23.9**	**25.8**	**22.1**	**24.2**	**n.a.**	**n.a.**
Belarus	3.0	<2.5	n.a.	n.a.	4.5	n.a.	9.7	n.a.	24.6	26.6	20.4	22.6	19.0	n.a.
Bulgaria	6.5	3.0	n.a.	n.a.	n.a.	n.a.	n.a.	n.a.	25.3	27.4	24.2	26.4	n.a.	n.a.
Czechia	<2.5	<2.5	n.a.	n.a.	n.a.	n.a.	n.a.	n.a.	26.6	28.5	23.3	25.7	n.a.	n.a.
Hungary	<2.5	<2.5	1.0	n.a.	n.a.	n.a.	n.a.	n.a.	26.4	28.6	23.6	25.8	n.a.	n.a.
Poland	<2.5	<2.5	1.1	n.a.	n.a.	n.a.	n.a.	n.a.	23.4	25.6	23.5	25.7	n.a.	n.a.
Republic of Moldova	n.a.	n.a.	3.4	n.a.	6.4	n.a.	4.9	n.a.	18.3	20.1	25.6	26.8	36.4	n.a.
Romania	<2.5	<2.5	4.1	n.a.	n.a.	n.a.	n.a.	n.a.	22.1	24.5	24.6	26.7	n.a.	n.a.
Russian Federation	<2.5	<2.5	n.a.	n.a.	n.a.	n.a.	n.a.	n.a.	23.9	25.7	21.3	23.3	n.a.	n.a.
Slovakia	6.2	2.7	<0.5	n.a.	n.a.	n.a.	n.a.	n.a.	20.4	22.4	24.5	26.6	n.a.	n.a.
Ukraine	<2.5	3.3	<0.5	n.a.	n.a.	n.a.	n.a.	n.a.	24.2	26.1	21.3	23.5	19.7	n.a.

TABLE A1.1
(CONTINUED)

REGIONS/SUBREGIONS/COUNTRIES	PREVALENCE OF UNDERNOURISHMENT IN THE TOTAL POPULATION[1] (%)		PREVALENCE OF SEVERE FOOD INSECURITY IN THE TOTAL POPULATION[1,2] (%)	PREVALENCE OF WASTING IN CHILDREN (UNDER 5 YEARS OF AGE) (%)	PREVALENCE OF STUNTING IN CHILDREN (UNDER 5 YEARS OF AGE) (%)		PREVALENCE OF OVERWEIGHT IN CHILDREN (UNDER 5 YEARS OF AGE) (%)		PREVALENCE OF OBESITY IN THE ADULT POPULATION (18 YEARS AND OLDER) (%)		PREVALENCE OF ANAEMIA AMONG WOMEN OF REPRODUCTIVE AGE (15–49) (%)		PREVALENCE OF EXCLUSIVE BREASTFEEDING AMONG INFANTS 0–5 MONTHS OF AGE (%)	
	2004–06	2015–17	2015–17	2017[3]	2012[4]	2017[3]	2012[4]	2017[3]	2012	2016	2012	2016[5]	2012[6]	2017[7]
Northern Europe	**<2.5**	**<2.5**	**2.6**	**n.a.**	**n.a.**	**n.a.**	**n.a.**	**n.a.**	**25.2**	**27.5**	**12.6**	**16.0**	**n.a.**	**n.a.**
Denmark	<2.5	<2.5	1.0	n.a.	n.a.	n.a.	n.a.	n.a.	19.8	21.3	13.4	16.3	n.a.	n.a.
Estonia	4.2	2.8	<0.5	n.a.	n.a.	n.a.	n.a.	n.a.	22.3	23.8	23.4	25.6	n.a.	n.a.
Finland	<2.5	<2.5		n.a.	n.a.	n.a.	n.a.	n.a.	23.2	24.9	13.1	15.9	n.a.	n.a.
Iceland	<2.5	<2.5	1.6	n.a.	n.a.	n.a.	n.a.	n.a.	21.3	23.1	13.2	16.1	n.a.	n.a.
Ireland	<2.5	<2.5	2.7[d]	n.a.	n.a.	n.a.	n.a.	n.a.	23.9	26.9	12.2	14.8	n.a.	n.a.
Latvia	<2.5	<2.5	0.7	n.a.	n.a.	n.a.	n.a.	n.a.	24.2	25.7	22.9	25.1	n.a.	n.a.
Lithuania	<2.5	<2.5	<0.5	n.a.	n.a.	n.a.	n.a.	n.a.	26.7	28.4	23.2	25.5	n.a.	n.a.
Norway	<2.5	<2.5	1.2	n.a.	n.a.	n.a.	n.a.	n.a.	23.0	25.0	12.7	15.3	n.a.	n.a.
Sweden	<2.5	<2.5	1.0	n.a.	n.a.	n.a.	n.a.	n.a.	20.4	22.1	12.8	15.4	n.a.	n.a.
United Kingdom	<2.5	<2.5	3.4	n.a.	n.a.	n.a.	n.a.	n.a.	26.9	29.5	11.5	15.3	n.a.	n.a.
Southern Europe	**<2.5**	**<2.5**	**1.7**	**n.a.**	**n.a.**	**n.a.**	**n.a.**	**n.a.**	**22.8**	**24.6**	**15.8**	**18.6**	**n.a.**	**n.a.**
Albania	10.9	5.5	10.5	n.a.	23.1	n.a.	23.4	n.a.	19.9	22.3	22.7	25.3	37.1	n.a.
Andorra	n.a.	n.a.	n.a.	n.a.	n.a.	n.a.	n.a.	n.a.	26.6	28.0	11.6	13.9	n.a.	n.a.
Bosnia and Herzegovina	3.2	<2.5	1.5	n.a.	8.9	n.a.	17.4	n.a.	17.7	19.4	27.1	29.4	18.2	n.a.
Croatia	2.9	<2.5	0.8	n.a.	n.a.	n.a.	n.a.	n.a.	24.9	27.1	25.2	27.3	n.a.	n.a.
Greece	<2.5	<2.5	3.1	n.a.	n.a.	n.a.	n.a.	n.a.	25.4	27.4	13.1	15.9	n.a.	n.a.
Italy	<2.5	<2.5	1.0	n.a.	n.a.	n.a.	n.a.	n.a.	21.4	22.9	14.4	17.3	n.a.	n.a.
Malta	<2.5	<2.5		n.a.	n.a.	n.a.	n.a.	n.a.	29.5	31.0	13.7	16.4	n.a.	n.a.
Montenegro	--	<2.5	2.3	2.8	7.9	9.4	15.6	22.3	23.1	24.9	22.8	25.2	19.3	16.8
Portugal	<2.5	<2.5	3.7	n.a.	n.a.	n.a.	n.a.	n.a.	21.0	23.2	14.7	17.5	n.a.	n.a.
Serbia	--	5.6	2.1	3.9	6.6	6.0	15.6	13.9	21.6	23.5	24.9	27.2	13.4	12.8
Slovenia	<2.5	<2.5		n.a.	n.a.	n.a.	n.a.	n.a.	20.6	22.5	21.9	24.4	n.a.	n.a.

TABLE A1.1
(CONTINUED)

REGIONS/SUBREGIONS/COUNTRIES	PREVALENCE OF UNDERNOURISHMENT IN THE TOTAL POPULATION[1] (%)		PREVALENCE OF SEVERE FOOD INSECURITY IN THE TOTAL POPULATION[1,2] (%)	PREVALENCE OF WASTING IN CHILDREN (UNDER 5 YEARS OF AGE) (%)	PREVALENCE OF STUNTING IN CHILDREN (UNDER 5 YEARS OF AGE) (%)		PREVALENCE OF OVERWEIGHT IN CHILDREN (UNDER 5 YEARS OF AGE) (%)		PREVALENCE OF OBESITY IN THE ADULT POPULATION (18 YEARS AND OLDER) (%)		PREVALENCE OF ANAEMIA AMONG WOMEN OF REPRODUCTIVE AGE (15–49) (%)		PREVALENCE OF EXCLUSIVE BREASTFEEDING AMONG INFANTS 0–5 MONTHS OF AGE (%)	
	2004–06	2015–17	2015–17	2017[3]	2012[4]	2017[3]	2012[4]	2017[3]	2012	2016	2012	2016[5]	2012[6]	2017[7]
Spain	<2.5	<2.5	1.4	n.a.	n.a.	n.a.	n.a.	n.a.	25.0	27.1	13.8	16.6	n.a.	n.a.
The former Yugoslav Republic of Macedonia	6.1	4.1	3.3	n.a.	4.9	n.a.	12.4	n.a.	21.9	23.9	19.5	23.3	23.0	n.a.
Western Europe	**<2.5**	**<2.5**	**1.1**	**n.a.**	**n.a.**	**n.a.**	**n.a.**	**n.a.**	**22.4**	**24.2**	**14.0**	**17.0**	**n.a.**	**n.a.**
Austria	<2.5	<2.5		n.a.	n.a.	n.a.	n.a.	n.a.	20.1	21.9	14.4	17.3	n.a.	n.a.
Belgium	<2.5	<2.5		n.a.	n.a.	n.a.	n.a.	n.a.	22.9	24.5	13.4	16.2	n.a.	n.a.
France	<2.5	<2.5	1.1	n.a.	n.a.	n.a.	n.a.	n.a.	21.6	23.2	14.9	18.1	n.a.	n.a.
Germany	<2.5	<2.5	0.8	n.a.	1.3	n.a.	3.5	n.a.	23.7	25.7	13.4	16.3	n.a.	n.a.
Luxembourg	<2.5	<2.5	1.0	n.a.	n.a.	n.a.	n.a.	n.a.	22.4	24.2	13.3	16.1	n.a.	n.a.
Netherlands	<2.5	<2.5		n.a.	n.a.	n.a.	n.a.	n.a.	21.0	23.1	13.4	16.4	n.a.	n.a.
Switzerland	<2.5	<2.5	1.3	n.a.	n.a.	n.a.	n.a.	n.a.	19.6	21.2	15.1	18.3	n.a.	n.a.

[1] Regional and subregional estimates were included when more than 50 percent of population was covered. To reduce the margin of error, estimates are presented as three-year averages.

[2] FAO estimate of the percentage of people in the total population living in households where at least one adult has been found to be food insecure. To reduce the impact of year-to-year sampling variability, estimates are presented as three-year averages. Country-level results are presented only for those countries for which estimates are based on official national data (Ecuador, Ghana, Malawi, Republic of Korea, Saint Lucia, Seychelles and the United States of America) or as provisional estimates, based on FAO Voices of the Hungry data collected through the Gallup World Poll, for countries whose national statistical authorities (NSAs) provided permission to publish them. Note that consent to publication does not necessarily imply validation of the estimate by the NSAs and that the estimate is subject to revision as soon as suitable data from official national sources are available. Global, regional and subregional aggregates reflect data collected in approximately 150 countries.

[3] For regional estimates, values correspond to the model predicted estimate for the year 2017. For countries, the latest data available from 2013 to 2017 are used.

[4] For regional estimates, values correspond to the model predicted estimate for the year 2012. For countries, the latest data available from 2005 to 2012 are used.

[5] Anaemia data for 2016 for countries in the WHO European region are undergoing validation and thus are subject to change. The WHO European region includes: Albania, Andorra, Armenia, Austria, Azerbaijan, Belarus, Belgium, Bosnia and Herzegovina, Bulgaria, Croatia, Cyprus, Czechia, Denmark, Estonia, Finland, France, Georgia, Germany, Greece, Hungary, Iceland, Ireland, Israel, Italy, Kazakhstan, Kyrgyzstan, Latvia, Lithuania, Luxembourg, Malta, Monaco, Montenegro, Netherlands, Norway, Poland, Portugal, Republic of Moldova, Romania, Russian Federation, San Marino, Serbia, Slovakia, Slovenia, Spain, Sweden, Switzerland, Tajikistan, The former Yugoslav Republic of Macedonia, Turkey, Turkmenistan, Ukraine, United Kingdom of Great Britain and Northern Ireland, and Uzbekistan.

[6] Regional estimates are included when more than 50 percent of population is covered. For countries, the latest data available from 2005 to 2012 are used.

[7] Regional estimates are included when more than 50 percent of population is covered. For countries, the latest data available from 2013 to 2018 are used.

* Wasting, stunting and overweight under 5 years of age, and exclusive breastfeeding regional aggregates exclude Japan.

[a] The Central Agency for Public Mobilization & Statistics (CAPMAS) reports an estimate of severe food insecurity of 1.3 percent for 2015, based on HIECS data, using the WFP consolidated approach for reporting indicators of food security. Note that the two estimates are not directly comparable due to different definitions of "severe food insecurity".

[b] Consecutive low population coverage; interpret with caution.

[c] Based on a combination of official national data and FAO data.

[d] The Government of Ireland reports estimates of the "Proportion of the population at risk of food poverty" produced by the Central Statistics Office (CSO) and Economic and Social Research Institute (ESRI) as part of the Survey on Income and Social Conditions (SILC) 2015, as a proxy for SDG indicator 2.1.2. See http://irelandsdg.geohive.ie/datasets/sdg-2.1.2-prevalence-of-moderate-or-severe-food-insecurity-in-the-population-based-on-the-food-insecurity-experience-scale-nuts-3-2015-ireland-cso-amp-osi

<2.5 = proportion of undernourishment less than 2.5 percent; <0.5 = prevalence of severe food insecurity less than 0.5 percent.

n.a. = data not available.

TABLE A1.2
PROGRESS TOWARDS THE SUSTAINABLE DEVELOPMENT GOALS (SDGs): NUMBER OF PEOPLE WHO ARE AFFECTED BY UNDERNOURISHMENT, SEVERE FOOD INSECURITY AND SELECTED FORMS OF MALNUTRITION; NUMBER OF INFANTS EXCLUSIVELY BREASTFED

REGIONS/SUBREGIONS/COUNTRIES	NUMBER OF UNDERNOURISHED PEOPLE[1] (millions)		NUMBER OF SEVERELY FOOD-INSECURE PEOPLE[1,2] (millions)	NUMBER OF CHILDREN (UNDER 5 YEARS OF AGE) AFFECTED BY WASTING[3] (millions)	NUMBER OF CHILDREN (UNDER 5 YEARS OF AGE) WHO ARE STUNTED[3] (millions)		NUMBER OF CHILDREN (UNDER 5 YEARS OF AGE) WHO ARE OVERWEIGHT[3] (millions)		NUMBER OF ADULTS (18 YEARS AND OLDER) WHO ARE OBESE (millions)		NUMBER OF WOMEN OF REPRODUCTIVE AGE (15-49) AFFECTED BY ANAEMIA[5] (millions)		NUMBER OF INFANTS 0–5 MONTHS OF AGE EXCLUSIVELY BREASTFED (millions)	
	2004–06	2015–17	2015–17	2017[3]	2012[4]	2017[3]	2012[4]	2017[3]	2012	2016	2012	2016[5]	2012[6]	2017[7]
WORLD	938.4	803.1	684.7	50.5	165.2	150.8	35.7	38.3	563.7	672.3	552.2	613.2	49.7	55.4
Least developed countries	215.7	237.1	231.0	12.4	49.8	48.9	4.5	5.8	20.5	28.3	85.0	95.3	12.6	15.3
Landlocked developing countries	97.6	110.0	99.2	5.0	23.9	23.1	2.7	2.9	17.0	21.9	34.6	39.7	6.5	8.0
Small island developing states	12.5	11.5	n.a.	n.a.	n.a.	n.a.	n.a.	n.a.	7.3	8.7	4.9	5.3	0.4	0.4
Low-income economies	151.7	185.6	n.a.	7.9	37.8	37.8	3.2	3.4	13.8	18.8	51.3	57.9	9.0	11.5
Lower-middle-income economies	487.4	423.5	n.a.	37.0	112.1	101.1	11.8	12.5	106.8	137.6	304.1	328.2	25.4	29.9
Low-income food-deficit countries	518.9	518.0	n.a.	39.5	126.7	116.6	10.3	10.6	<0.1	<0.1	300.1	325.3	27.5	33.1
AFRICA	196.5	239.9	317.7	13.8	57.2	58.7	8.7	9.7	58.5	73.5	98.9	109.8	13.3	17.5
Northern Africa	9.6	19.2	26.1	2.3	4.9	5.0	2.5	3.0	29.1	35.5	17.2	18.6	2.2	2.6
Algeria	2.9	1.9	n.a.	n.a.	0.5	n.a.	0.5	n.a.	6.0	7.4	3.5	3.8	0.2	n.a.
Egypt	4.2	4.6	9.7[a]	1.1	2.9	2.7	1.9	1.9	14.2	17.1	6.5	6.7	1.3	1.0
Libya	n.a.	n.a.	n.a.	n.a.	0.1	n.a.	0.1	n.a.	1.1	1.3	0.5	0.6	n.a.	n.a.
Morocco	1.7	1.4	n.a.	n.a.	0.5	n.a.	0.3	n.a.	4.8	5.9	3.1	3.5	0.2	n.a.
Sudan	--	10.0	n.a.	0.9	1.9	2.2	0.1	0.2	1.1	1.6	2.7	3.1	0.5	0.7
Tunisia	0.6	0.6	n.a.	n.a.	0.1	n.a.	0.1	n.a.	1.9	2.3	0.9	1.0	<0.1	n.a.
Northern Africa (excluding Sudan)	9.6	9.2	16.6	n.a.	n.a.	n.a.	n.a.	n.a.	28.1	33.9	14.5	15.5	1.8	1.8
Sub-Saharan Africa	177.3	220.7	291.6	11.4	52.3	53.8	6.2	6.8	30.5	39.5	81.8	91.2	11.0	15.0
Eastern Africa	113.7	127.7	120.1	4.0	23.6	23.9	2.8	3.0	7.7	10.7	25.9	30.1	6.3	8.3
Burundi	n.a.	n.a.	n.a.	0.1	0.9	1.1	<0.1	<0.1	0.2	0.2	0.6	0.7	0.3	0.3
Comoros	n.a.	n.a.	n.a.	n.a.	<0.1	n.a.	<0.1	n.a.	<0.1	<0.1	0.1	0.1	<0.1	n.a.
Djibouti	0.3	0.2	n.a.	0.1	<0.1	n.a.	<0.1	n.a.	0.1	0.1	0.1	0.1	<0.1	n.a.

TABLE A1.2
(CONTINUED)

REGIONS/SUBREGIONS/COUNTRIES	NUMBER OF UNDERNOURISHED PEOPLE[1] (millions)		NUMBER OF SEVERELY FOOD-INSECURE PEOPLE[1,2] (millions)	NUMBER OF CHILDREN (UNDER 5 YEARS OF AGE) AFFECTED BY WASTING (millions)	NUMBER OF CHILDREN (UNDER 5 YEARS OF AGE) WHO ARE STUNTED (millions)		NUMBER OF CHILDREN (UNDER 5 YEARS OF AGE) WHO ARE OVERWEIGHT (millions)		NUMBER OF ADULTS (18 YEARS AND OLDER) WHO ARE OBESE (millions)		NUMBER OF WOMEN OF REPRODUCTIVE AGE (15-49) AFFECTED BY ANAEMIA (millions)		NUMBER OF INFANTS 0-5 MONTHS OF AGE EXCLUSIVELY BREASTFED (millions)	
	2004-06	2015-17	2015-17	2017[3]	2012[4]	2017[3]	2012[4]	2017[3]	2012	2016	2012	2016[5]	2012[6]	2017[7]
Eritrea	n.a.	n.a.	n.a.	n.a.	0.4	n.a.	<0.1	n.a.	0.1	0.1	0.4	0.5	0.1	n.a.
Ethiopia	30.5	21.9		1.5	6.2	5.8	0.3	0.4	1.3	1.9	4.7	5.8	1.5	1.8
Kenya	10.2	11.7	17.3	0.3	2.3	1.8	0.3	0.3	1.1	1.5	2.8	3.1	0.4	0.9
Madagascar	6.4	10.7	n.a.	n.a.	1.6	n.a.	n.a.	n.a.	0.4	0.6	1.9	2.2	0.3	n.a.
Malawi	3.4	4.8	9.5	0.1	1.3	1.1	0.3	0.1	0.3	0.4	1.1	1.4	0.4	0.4
Mauritius	<0.1	<0.1	<0.1	n.a.	n.a.	n.a.	n.a.	n.a.	0.1	0.1	0.1	0.1	n.a.	n.a.
Mozambique	7.7	8.8		n.a.	1.9	n.a.	0.4	n.a.	0.6	0.8	2.9	3.4	0.4	0.4
Rwanda	4.1	4.3		<0.1	0.7	0.6	0.1	0.1	0.2	0.3	0.5	0.7	0.3	0.3
Seychelles	n.a.	n.a.	<0.1	n.a.	<0.1	n.a.	<0.1	n.a.	<0.1	<0.1	<0.1	<0.1	n.a.	n.a.
Somalia	n.a.	n.a.	n.a.	n.a.	0.6	n.a.	0.1	n.a.	0.3	0.4	1.0	1.1	<0.1	n.a.
South Sudan	--	n.a.	n.a.	n.a.	0.5	n.a.	0.1	n.a.	n.a.	n.a.	0.8	1.0	0.2	n.a.
Uganda	6.9	17.2		0.3	2.4	2.2	0.4	0.3	1.4	1.9	2.3	2.6	0.9	1.1
United Republic of Tanzania	13.6	17.8		0.4	2.9	3.2	0.5	0.3	0.6	0.8	4.3	4.7	0.9	1.2
Zambia	6.2	7.4		0.2	1.1	1.1	0.2	0.2	0.4	0.5	1.0	1.3	0.3	0.4
Zimbabwe	5.5	7.5		0.1	0.7	0.7	0.1	0.1	0.8	1.1	1.1	1.2	0.2	0.2
Middle Africa	**36.3**	**40.2**	**62.8**	**2.1**	**8.8**	**9.3**	**1.2**	**1.4**	**3.5**	**4.8**	**14.2**	**15.5**	**1.6**	**2.3**
Angola	10.7	6.9		0.3	1.2	2.0	n.a.	0.2	0.5	0.7	2.4	2.7	n.a.	0.4
Cameroon	3.5	1.7	8.3	0.2	1.1	1.2	0.2	0.2	0.9	1.2	2.1	2.4	0.1	0.2
Central African Republic	1.6	2.8	n.a.	n.a.	0.3	n.a.	<0.1	n.a.	0.1	0.2	0.5	0.6	0.1	n.a.
Chad	3.9	5.7	n.a.	0.3	0.9	1.0	0.1	0.1	0.2	0.3	1.3	1.5	<0.1	<0.1
Congo	1.5	1.9		0.1	0.2	0.2	<0.1	<0.1	0.2	0.2	0.5	0.6	<0.1	0.1
Democratic Republic of the Congo	n.a.	n.a.	n.a.	1.1	5.3	5.7	0.6	0.6	1.5	2.0	7.0	7.4	1.0	1.5

REGIONS/SUBREGIONS/COUNTRIES	NUMBER OF UNDERNOURISHED PEOPLE[1] (millions)		NUMBER OF SEVERELY FOOD-INSECURE PEOPLE[1,2] (millions)	NUMBER OF CHILDREN (UNDER 5 YEARS OF AGE) AFFECTED BY WASTING (millions)	NUMBER OF CHILDREN (UNDER 5 YEARS OF AGE) WHO ARE STUNTED (millions)		NUMBER OF CHILDREN (UNDER 5 YEARS OF AGE) WHO ARE OVERWEIGHT (millions)		NUMBER OF ADULTS (18 YEARS AND OLDER) WHO ARE OBESE (millions)		NUMBER OF WOMEN OF REPRODUCTIVE AGE (15–49) AFFECTED BY ANAEMIA (millions)		NUMBER OF INFANTS 0–5 MONTHS OF AGE EXCLUSIVELY BREASTFED (millions)	
	2004–06	2015–17	2015–17	2017[3]	2012[4]	2017[3]	2012[4]	2017[3]	2012	2016	2012	2016[5]	2012[6]	2017[7]
Equatorial Guinea	n.a.	n.a.	n.a.	n.a.	<0.1	n.a.	<0.1	n.a.	<0.1	<0.1	0.1	0.1	<0.1	n.a.
Gabon	0.2	0.2	n.a.	n.a.	<0.1	n.a.	<0.1	n.a.	0.1	0.1	0.2	0.3	<0.1	n.a.
Sao Tome and Principe	<0.1	<0.1	n.a.	<0.1	<0.1	<0.1	<0.1	<0.1	<0.1	<0.1	<0.1	<0.1	<0.1	<0.1
Southern Africa	**3.6**	**5.2**	**17.6**	**0.3**	**2.0**	**2.0**	**0.8**	**0.9**	**8.9**	**10.2**	**4.2**	**4.4**	**n.a.**	**0.5**
Botswana	0.6	0.6	0.9	n.a.	0.1	n.a.	<0.1	n.a.	0.2	0.2	0.2	0.2	<0.1	n.a.
Eswatini	0.2	0.3	n.a.	<0.1	0.1	<0.1	<0.1	<0.1	0.1	0.1	0.1	0.1	<0.1	<0.1
Lesotho	0.2	0.3	1.1	<0.1	0.1	0.1	<0.1	<0.1	0.1	0.2	0.1	0.2	<0.1	<0.1
Namibia	0.5	0.6	1.0	<0.1	0.1	0.1	<0.1	<0.1	0.2	0.2	0.2	0.2	<0.1	<0.1
South Africa	2.1	3.4		0.1	1.5	1.6	0.9	0.8	8.3	9.5	3.7	3.8	n.a.	0.4
Western Africa	**33.2**	**47.6**	**91.1**	**5.1**	**17.9**	**18.6**	**1.5**	**1.5**	**10.3**	**13.8**	**37.4**	**41.2**	**2.7**	**4.1**
Benin	1.2	1.1	n.a.	0.1	0.6	0.6	0.2	<0.1	0.4	0.5	1.2	1.3	0.1	0.2
Burkina Faso	3.3	4.0	4.4	0.2	1.0	0.9	0.1	<0.1	0.3	0.4	1.9	2.1	0.2	0.3
Cabo Verde	<0.1	<0.1	n.a.	n.a.	n.a.	n.a.	n.a.	n.a.	<0.1	<0.1	<0.1	<0.1	<0.1	n.a.
Côte d'Ivoire	3.7	4.9	n.a.	0.2	1.0	0.8	0.1	0.1	0.8	1.0	2.5	2.9	0.1	0.2
Gambia	0.2	0.2	0.5	<0.1	0.1	0.1	<0.1	<0.1	0.1	0.1	0.2	0.3	<0.1	0.1
Ghana	2.0	1.7	2.2	0.2	0.8	0.7	0.1	0.1	1.2	1.5	3.2	3.3	0.4	0.4
Guinea	2.1	2.4	5.0	0.2	0.6	0.6	0.1	0.1	0.3	0.4	1.4	1.5	0.1	0.1
Guinea-Bissau	0.3	0.5	n.a.	<0.1	0.1	0.1	<0.1	<0.1	0.1	0.1	0.2	0.2	<0.1	0.1
Liberia	1.3	1.8	2.6	<0.1	0.3	0.2	<0.1	0.1	0.2	0.2	0.4	0.4	<0.1	0.1
Mali	1.4	1.1	n.a.	0.4	0.8	1.0	0.1	0.1	0.4	0.5	1.9	2.0	0.1	0.3
Mauritania	0.4	0.5		0.1	0.1	0.2	<0.1	<0.1	0.2	0.3	0.3	0.4	<0.1	0.1
Niger	2.1	3.0	7.7	0.4	1.6	1.8	0.1	n.a.	0.3	0.4	1.8	2.1	0.2	n.a.

TABLE A1.2 (CONTINUED)

REGIONS/SUBREGIONS/COUNTRIES	NUMBER OF UNDERNOURISHED PEOPLE[1] (millions)		NUMBER OF SEVERELY FOOD-INSECURE PEOPLE[1,2] (millions)	NUMBER OF CHILDREN (UNDER 5 YEARS OF AGE) AFFECTED BY WASTING (millions)	NUMBER OF CHILDREN (UNDER 5 YEARS OF AGE) WHO ARE STUNTED (millions)		NUMBER OF CHILDREN (UNDER 5 YEARS OF AGE) WHO ARE OVERWEIGHT (millions)		NUMBER OF ADULTS (18 YEARS AND OLDER) WHO ARE OBESE (millions)		NUMBER OF WOMEN OF REPRODUCTIVE AGE (15–49) AFFECTED BY ANAEMIA (millions)		NUMBER OF INFANTS 0–5 MONTHS OF AGE EXCLUSIVELY BREASTFED (millions)	
	2004–06	2015–17	2015–17	2017[3]	2012[4]	2017[3]	2012[4]	2017[3]	2012	2016	2012	2016[5]	2012[6]	2017
Nigeria	9.1	21.5	46.1	3.4	10.2	13.9	0.9	0.5	5.4	7.3	19.1	21.1	0.9	1.6
Senegal	2.4	1.7		0.2	0.4	0.4	<0.1	<0.1	0.4	0.6	1.8	1.9	0.2	0.2
Sierra Leone	2.1	1.9		0.1	0.5	0.4	0.1	0.1	0.2	0.3	0.7	0.8	0.1	0.1
Togo	1.5	1.2	2.3	0.1	0.3	0.3	<0.1	<0.1	0.2	0.3	0.8	0.9	0.1	0.1
Sub-Saharan Africa (including Sudan)	186.9	230.7	301.1	n.a.	n.a.	n.a.	n.a.	n.a.	31.5	41.2	84.4	94.3	11.5	15.7
ASIA*	679.3	514.5	297.1	35.0	98.4	83.6	16.3	17.5	175.7	228.7	377.7	419.9	28.6	29.1
Central Asia	6.5	4.2	1.8	0.3b	1.1	0.9b	0.7	0.8b	5.9	7.4	5.9	6.2	0.5	0.6
Kazakhstan	0.9	<0.4	0.3	0.1	0.2	0.2	0.2	0.2	2.1	2.5	1.4	1.4	0.1	0.1
Kyrgyzstan	0.5	0.4		<0.1	0.1	0.1	0.1	0.1	0.4	0.6	0.5	0.6	0.1	0.1
Tajikistan	n.a.	n.a.	0.7	n.a.	0.3	n.a.	0.1	n.a.	0.5	0.6	0.6	0.7	0.1	0.1
Turkmenistan	0.2	0.3	n.a.	<0.1	0.1	0.1	<0.1	<0.1	0.5	0.6	0.5	0.5	<0.1	0.1
Uzbekistan	3.9	2.3		n.a.	0.5	n.a.	0.3	n.a.	2.4	3.1	3.0	3.0	0.2	n.a.
Eastern Asia*	218.0	139.1	12.8	1.7	7.1	4.8	4.9	4.8	61.9	81.3	89.4	107.4	5.4	3.5
China	206.0	124.5		1.6	7.8	6.9	5.5		54.7	72.9	78.1	95.0	4.7	3.1
China, mainland	204.7	123.5	n.a.	n.a.	n.a.	n.a.	n.a.	n.a.	n.a.	n.a.	n.a.	n.a.	n.a.	n.a.
Taiwan Province of China	1.1	0.8	n.a.	n.a.	n.a.	n.a.	n.a.	n.a.	n.a.	n.a.	n.a.	n.a.	n.a.	n.a.
China, Hong Kong SAR	<0.2	<0.2	n.a.	n.a.	n.a.	n.a.	n.a.	n.a.	n.a.	n.a.	n.a.	n.a.	n.a.	n.a.
China, Macao SAR	<0.1	<0.1	n.a.	n.a.	n.a.	n.a.	n.a.	n.a.	n.a.	n.a.	n.a.	n.a.	n.a.	n.a.
Democratic People's Republic of Korea	8.4	11.0	n.a.	n.a.	0.5	n.a.	0.0	n.a.	1.1	1.3	2.0	2.2	0.2	n.a.
Japan	<3.2	<3.2	<0.6	n.a.	0.4	n.a.	0.1	n.a.	4.1	4.7	5.2	5.6	n.a.	n.a.
Mongolia	0.8	0.6	<0.1	<0.1	0.1	<0.1	0.1	<0.1	0.3	0.4	0.1	0.2	<0.1	<0.1
Republic of Korea	<1.2	<1.3	<0.3c	n.a.	0.1	n.a.	0.2	n.a.	1.7	2.0	2.4	2.8	n.a.	n.a.

TABLE A1.2 (CONTINUED)

REGIONS/SUBREGIONS/COUNTRIES	NUMBER OF UNDERNOURISHED PEOPLE[1] (millions)		NUMBER OF SEVERELY FOOD-INSECURE PEOPLE[1,2] (millions)	NUMBER OF CHILDREN (UNDER 5 YEARS OF AGE) AFFECTED BY WASTING (millions)	NUMBER OF CHILDREN (UNDER 5 YEARS OF AGE) WHO ARE STUNTED (millions)		NUMBER OF CHILDREN (UNDER 5 YEARS OF AGE) WHO ARE OVERWEIGHT (millions)		NUMBER OF ADULTS (18 YEARS AND OLDER) WHO ARE OBESE (millions)		NUMBER OF WOMEN OF REPRODUCTIVE AGE (15–49) AFFECTED BY ANAEMIA (millions)		NUMBER OF INFANTS 0–5 MONTHS OF AGE EXCLUSIVELY BREASTFED (millions)	
	2004–06	2015–17	2015–17	2017[3]	2012[4]	2017[3]	2012[4]	2017[3]	2012	2016	2012	2016[5]	2012[6]	2017[7]
Eastern Asia (excluding China, mainland)	12.2	14.8	<1.2	n.a.	n.a.	n.a.	n.a.	n.a.	<0.1	<0.1	<0.1	<0.1	n.a.	n.a.
South-eastern Asia	101.4	62.2	55.9	5.1	16.6	14.9	3.2	4.2	21.7	29.4	43.1	48.5	3.9	18.5
Brunei Darussalam	<0.1	<0.1	n.a.	n.a.	<0.1	n.a.	<0.1	n.a.	<0.1	<0.1	<0.1	<0.1	n.a.	n.a.
Cambodia	2.7	2.9	2.3	0.2	0.7	0.6	<0.1	<0.1	0.3	0.4	1.9	2.0	0.3	0.2
Indonesia	41.9	20.2	n.a.	3.3	9.3	8.8	2.9	2.8	8.7	12.0	17.7	20.2	2.0	n.a.
Lao People's Democratic Republic	1.6	1.1	n.a.	n.a.	0.3	n.a.	<0.1	n.a.	0.1	0.2	0.6	0.7	0.1	n.a.
Malaysia	1.0	0.9	n.a.	0.3	0.4	0.5	0.4	0.2	2.5	3.3	1.8	2.1	n.a.	n.a.
Myanmar	15.5	5.6	5.5	0.3	1.8	1.3	0.1	0.1	1.6	2.2	6.0	6.9	0.2	0.5
Philippines	14.1	14.2	13.3	0.8	3.7	3.8	0.5	0.4	2.9	3.8	4.5	4.2	0.8	n.a.
Singapore	n.a.	n.a.	0.2	n.a.	n.a.	n.a.	n.a.	n.a.	0.3	0.3	0.3	0.3	n.a.	n.a.
Thailand	8.2	6.2	n.a.	0.2	0.7	0.4	0.4	0.3	4.3	5.8	4.8	5.6	0.1	0.2
Timor-Leste	0.3	0.3	n.a.	<0.1	0.1	0.1	<0.1	<0.1	<0.1	<0.1	0.1	0.1	<0.1	<0.1
Viet Nam	15.3	10.2	2.2	0.5	1.7	1.9	0.3	0.4	1.0	1.5	5.4	6.3	0.3	0.4
Southern Asia	334.0	279.9	201.2	26.9	69.0	58.7	5.3	5.4	46.9	62.5	218.5	234.2	16.9	18.5
Afghanistan	8.3	10.5	5.5	0.5	n.a.	2.1	n.a.	0.3	0.5	0.7	2.4	3.2	n.a.	0.5
Bangladesh	23.8	24.8		2.2	6.5	5.5	0.2	0.2	2.5	3.6	17.4	18.2	1.7	1.7
Bhutan	n.a.	n.a.	n.a.	n.a.	<0.1	n.a.	<0.1	n.a.	<0.1	<0.1	0.1	0.1	<0.1	<0.1
India	253.9	195.9		25.5	62.2	46.6	2.5	2.5	24.1	32.8	165.6	175.6	11.4	13.2
Iran (Islamic Republic of)	4.3	4.0		n.a.	0.5	n.a.	<0.1	n.a.	12.0	14.7	6.4	7.2	0.7	n.a.
Maldives	<0.1	<0.1	n.a.	n.a.	<0.1	n.a.	<0.1	n.a.	<0.1	<0.1	<0.1	<0.1	<0.1	n.a.
Nepal	4.1	2.8	2.3	0.3	1.3	1.0	0.5	<0.1	0.5	0.7	2.6	2.8	0.4	0.4
Pakistan	35.9	39.5		n.a.	10.7	n.a.	1.1	n.a.	6.7	9.1	22.4	25.3	1.9	1.9
Sri Lanka	3.6	2.3		0.2	0.3	0.3	<0.1	<0.1	0.6	0.8	1.6	1.7	0.3	0.3

TABLE A1.2 (CONTINUED)

REGIONS/SUBREGIONS/COUNTRIES	NUMBER OF UNDERNOURISHED PEOPLE[1] (millions)		NUMBER OF SEVERELY FOOD-INSECURE PEOPLE[1,2] (millions)	NUMBER OF CHILDREN (UNDER 5 YEARS OF AGE) AFFECTED BY WASTING (millions)	NUMBER OF CHILDREN (UNDER 5 YEARS OF AGE) WHO ARE STUNTED (millions)		NUMBER OF CHILDREN (UNDER 5 YEARS OF AGE) WHO ARE OVERWEIGHT (millions)		NUMBER OF ADULTS (18 YEARS AND OLDER) WHO ARE OBESE (millions)		NUMBER OF WOMEN OF REPRODUCTIVE AGE (15–49) AFFECTED BY ANAEMIA (millions)		NUMBER OF INFANTS 0–5 MONTHS OF AGE EXCLUSIVELY BREASTFED (millions)	
	2004–06	2015–17	2015–17	2017[3]	2012[4]	2017[3]	2012[4]	2017[3]	2012	2016	2012	2016[5]	2012[6]	2017
Southern Asia (excluding India)	**80.2**	**84.0**	**44.9**	**n.a.**	**n.a.**	**n.a.**	**n.a.**	**n.a.**	**n.a.**	**n.a.**	**n.a.**	**n.a.**	**5.5**	**5.3**
Western Asia	**19.5**	**29.1**	**25.3**	**1.1**	**4.6**	**4.2**	**2.1**	**2.3**	**39.3**	**48.0**	**20.8**	**23.7**	**n.a.**	**n.a.**
Armenia	0.2	0.1	0.1	<0.1	<0.1	<0.1	<0.1	<0.1	0.4	0.5	0.2	0.2	<0.1	<0.1
Azerbaijan	0.5	<0.2		<0.1	0.1	0.1	0.1	0.1	1.2	1.4	1.0	1.0	<0.1	<0.1
Bahrain	n.a.	n.a.		n.a.	n.a.	n.a.	n.a.	n.a.	0.3	0.3	0.1	0.1	n.a.	n.a.
Cyprus	<0.1	<0.1		n.a.	n.a.	n.a.	n.a.	n.a.	0.2	0.2	0.1	0.1	n.a.	n.a.
Georgia	0.3	0.3	0.3	n.a.	<0.1	n.a.	0.1	n.a.	0.7	0.8	0.3	0.3	<0.1	n.a.
Iraq	7.6	10.3		n.a.	1.1	n.a.	0.6	n.a.	4.4	5.5	2.3	2.7	0.2	n.a.
Israel	<0.2	<0.2		n.a.	n.a.	n.a.	n.a.	n.a.	1.3	1.4	0.2	0.3	n.a.	n.a.
Jordan	0.4	1.3	1.3	n.a.	0.1	n.a.	0.1	n.a.	1.3	1.6	0.6	0.7	0.1	n.a.
Kuwait	<0.1	<0.1		<0.1	<0.1	<0.1	<0.1	<0.1	0.8	1.0	0.2	0.3	n.a.	n.a.
Lebanon	0.1	0.7	n.a.	n.a.	n.a.	n.a.	n.a.	n.a.	1.0	1.2	0.4	0.5	n.a.	n.a.
Oman	0.3	0.2	n.a.	<0.1	<0.1	0.1	<0.1	<0.1	0.5	0.7	0.3	0.3	n.a.	<0.1
Palestine	n.a.	n.a.	0.4	n.a.	n.a.	n.a.	n.a.	n.a.	n.a.	n.a.	n.a.	n.a.	n.a.	n.a.
Qatar	n.a.	n.a.	n.a.	n.a.	n.a.	n.a.	n.a.	n.a.	0.5	0.7	0.1	0.1	<0.1	n.a.
Saudi Arabia	1.9	1.8		n.a.	0.3	n.a.	0.2	n.a.	5.9	7.2	3.0	3.4	n.a.	n.a.
Syrian Arab Republic	n.a.	n.a.	n.a.	n.a.	0.7	n.a.	0.5	n.a.	2.9	3.5	1.6	1.6	0.2	n.a.
Turkey	<1.7	<2		0.1	0.8	0.6	n.a.	0.7	14.8	17.6	5.9	6.5	n.a.	0.4
United Arab Emirates	0.2	0.2		n.a.	n.a.	n.a.	n.a.	n.a.	1.8	2.4	0.4	0.5	n.a.	n.a.
Yemen	6.2	9.5	n.a.	0.6	1.7	1.8	0.1	0.1	1.5	2.0	4.0	4.8	n.a.	0.1

TABLE A1.2
(CONTINUED)

REGIONS/SUBREGIONS/COUNTRIES	NUMBER OF UNDERNOURISHED PEOPLE[1] (millions)		NUMBER OF SEVERELY FOOD-INSECURE PEOPLE[1,2] (millions)	NUMBER OF CHILDREN (UNDER 5 YEARS OF AGE) AFFECTED BY WASTING (millions)	NUMBER OF CHILDREN (UNDER 5 YEARS OF AGE) WHO ARE STUNTED (millions)		NUMBER OF CHILDREN (UNDER 5 YEARS OF AGE) WHO ARE OVERWEIGHT (millions)		NUMBER OF ADULTS (18 YEARS AND OLDER) WHO ARE OBESE (millions)		NUMBER OF WOMEN OF REPRODUCTIVE AGE (15–49) AFFECTED BY ANAEMIA (millions)		NUMBER OF INFANTS 0–5 MONTHS OF AGE EXCLUSIVELY BREASTFED (millions)	
	2004–06	2015–17	2015–17	2017[3]	2012[4]	2017[3]	2012[4]	2017[3]	2012	2016	2012	2016[5]	2012[6]	2017[7]
Central Asia and Southern Asia	340.5	284.1	203.0	27.2	70.1	59.6	6.0	6.2	52.8	69.9	224.4	240.4	17.4	19.2
Eastern Asia and South-eastern Asia*	319.4	201.3	68.7	6.7	23.6	19.7	8.2	9.0	83.6	110.7	132.5	155.9	9.3	6.5
Western Asia and Northern Africa	29.1	48.3	51.4	3.4	9.5	9.2	4.5	5.2	68.4	83.6	38.0	42.3	4.0	4.1
LATIN AMERICA AND THE CARIBBEAN	51.0	39.0	n.a.	0.7	6.1	5.1	3.8	3.9	88.3	104.7	34.9	37.6	3.6	n.a.
Caribbean	9.1	7.2	n.a.	0.1^b	0.4	0.3^b	0.2	0.3^b	5.5	6.6	3.2	3.4	0.2	0.2
Antigua and Barbuda	n.a.	n.a.	n.a.	n.a.	n.a.	n.a.	n.a.	n.a.	<0.1	<0.1	<0.1	<0.1	n.a.	n.a.
Bahamas	n.a.	n.a.	n.a.	n.a.	n.a.	n.a.	n.a.	n.a.	0.1	0.1	<0.1	<0.1	n.a.	n.a.
Barbados	<0.1	<0.1	n.a.	n.a.	<0.1	n.a.	<0.1	n.a.	<0.1	0.1	<0.1	<0.1	<0.1	n.a.
Cuba	<0.3	<0.3	n.a.	n.a.	n.a.	n.a.	n.a.	n.a.	2.2	2.4	0.7	0.7	0.1	<0.1
Dominica	<0.1	<0.1	n.a.	n.a.	n.a.	n.a.	n.a.	n.a.	<0.1	<0.1	<0.1	<0.1	n.a.	n.a.
Dominican Republic	2.3	1.1		<0.1	0.1	0.1	0.1	0.1	1.5	1.9	0.8	0.8	<0.1	<0.1
Grenada	n.a.	n.a.	n.a.	n.a.	n.a.	n.a.	n.a.	n.a.	<0.1	<0.1	<0.1	<0.1	n.a.	n.a.
Haiti	5.3	5.0	n.a.	n.a.	0.3	n.a.	<0.1	n.a.	1.0	1.3	1.3	1.3	0.1	0.1
Jamaica	0.2	0.3	n.a.	<0.1	<0.1	<0.1	<0.1	<0.1	0.4	0.5	0.2	0.2	<0.1	n.a.
Puerto Rico	n.a.	n.a.	n.a.	n.a.	n.a.	n.a.	n.a.	n.a.	n.a.	n.a.	n.a.	n.a.	n.a.	n.a.
Saint Kitts and Nevis	n.a.	n.a.	n.a.	n.a.	n.a.	n.a.	n.a.	n.a.	<0.1	<0.1	<0.1	<0.1	n.a.	n.a.
Saint Lucia	<0.1	<0.1	<0.1	n.a.	<0.1	n.a.	<0.1	n.a.	<0.1	<0.1	<0.1	<0.1	n.a.	n.a.
Saint Vincent and the Grenadines	<0.1	<0.1	n.a.	n.a.	<0.1	n.a.	<0.1	n.a.	<0.1	<0.1	<0.1	<0.1	n.a.	n.a.
Trinidad and Tobago	0.2	0.2	n.a.	n.a.	<0.1	n.a.	<0.1	n.a.	0.2	0.2	0.1	0.1	<0.1	n.a.
Central America	12.3	11.1	18.1	0.1	2.7	2.3	1.0	1.0	25.4	30.4	6.9	7.4	0.7	1.1
Belize	<0.1	<0.1	n.a.	<0.1	<0.1	<0.1	<0.1	<0.1	<0.1	<0.1	<0.1	<0.1	<0.1	<0.1
Costa Rica	0.2	0.2	0.2	n.a.	<0.1	n.a.	<0.1	n.a.	0.8	0.9	0.2	0.2	<0.1	<0.1
El Salvador	0.6	0.7	0.7	<0.1	0.1	0.1	<0.1	<0.1	0.8	1.0	0.3	0.4	<0.1	0.1

REGIONS/SUBREGIONS/COUNTRIES	NUMBER OF UNDERNOURISHED PEOPLE[1] (millions)		NUMBER OF SEVERELY FOOD-INSECURE PEOPLE[1,2] (millions)	NUMBER OF CHILDREN (UNDER 5 YEARS OF AGE) AFFECTED BY WASTING (millions)	NUMBER OF CHILDREN (UNDER 5 YEARS OF AGE) WHO ARE STUNTED (millions)		NUMBER OF CHILDREN (UNDER 5 YEARS OF AGE) WHO ARE OVERWEIGHT (millions)		NUMBER OF ADULTS (18 YEARS AND OLDER) WHO ARE OBESE (millions)		NUMBER OF WOMEN OF REPRODUCTIVE AGE (15–49) AFFECTED BY ANAEMIA (millions)		NUMBER OF INFANTS 0–5 MONTHS OF AGE EXCLUSIVELY BREASTFED (millions)	
	2004–06	2015–17	2015–17	2017[3]	2012[4]	2017[3]	2012[4]	2017[3]	2012	2016	2012	2016[5]	2012[6]	2017[7]
Guatemala	2.1	2.6		<0.1	0.9	0.9	0.1	0.1	1.3	1.7	0.7	0.7	0.2	0.2
Honduras	1.3	1.4		n.a.	0.2	n.a.	0.1	n.a.	0.8	1.0	0.3	0.4	0.1	n.a.
Mexico	6.0	4.8	11.3	0.1	1.6	1.4	1.0	0.6	20.5	24.3	4.9	5.1	0.3	0.7
Nicaragua	1.3	1.0		n.a.	0.1	n.a.	0.1	n.a.	0.7	0.9	0.2	0.3	<0.1	n.a.
Panama	0.8	0.4		n.a.	0.1	n.a.	n.a.	n.a.	0.5	0.6	0.2	0.2	n.a.	<0.1
South America	**29.6**	**20.7**	**29.0**	**0.4**b	**3.0**	**2.5**b	**2.6**	**2.6**b	**57.4**	**67.7**	**24.8**	**26.9**	**2.8**	**n.a.**
Argentina	1.9	1.7	3.8	n.a.	0.3	n.a.	0.4	n.a.	7.6	8.7	1.7	2.0	0.2	n.a.
Bolivia (Plurinational State of)	2.8	2.2	1.2c	<0.1	0.2	0.2	0.1	0.1	1.0	1.3	0.8	0.8	0.2	0.1
Brazil	8.6	<5.2	n.a.	n.a.	1.1	n.a.	1.1	n.a.	27.8	33.1	14.1	15.5	1.2	n.a.
Chile	0.6	0.6	0.8	<0.1	<0.1	<0.1	0.1	0.1	3.4	3.9	0.5	0.7	n.a.	n.a.
Colombia	4.2	3.2	n.a.	n.a.	0.5	n.a.	0.2	n.a.	6.3	7.5	2.9	2.8	n.a.	n.a.
Ecuador	2.3	1.3	n.a.	<0.1	0.4	0.4	0.1	0.1	1.7	2.1	0.7	0.8	n.a.	n.a.
Guyana	<0.1	<0.1		<0.1	<0.1	<0.1	<0.1	<0.1	0.1	0.1	0.1	0.1	<0.1	<0.1
Paraguay	0.7	0.8	n.a.	<0.1	0.1	<0.1	0.1	0.1	0.7	0.9	0.3	0.4	<0.1	<0.1
Peru	5.4	2.8	n.a.	<0.1	0.5	0.4	0.2	n.a.	3.4	4.0	1.6	1.6	0.4	0.4
Suriname	<0.1	<0.1		n.a.	<0.1	n.a.	<0.1	n.a.	0.1	0.1	<0.1	<0.1	<0.1	n.a.
Uruguay	0.1	<0.1		n.a.	<0.1	n.a.	<0.1	n.a.	0.7	0.7	0.2	0.2	n.a.	n.a.
Venezuela (Bolivarian Republic of)	2.8	3.7	n.a.	n.a.	0.4	n.a.	0.2	n.a.	4.6	5.4	1.8	2.0	n.a.	n.a.
OCEANIA	**1.8**	**2.6**	**n.a.**	**n.a.**	**n.a.**	**n.a.**	**n.a.**	**n.a.**	**7.0**	**8.1**	**1.3**	**1.5**	**n.a.**	**n.a.**
Australia and New Zealand	**<0.6**	**<0.7**	**0.9**	**n.a.**	**n.a.**	**n.a.**	**n.a.**	**n.a.**	**6.0**	**6.8**	**0.6**	**0.6**	**n.a.**	**n.a.**
Australia	<0.5	<0.6	0.8	n.a.	<0.1	n.a.	0.1	n.a.	5.0	5.7	0.4	0.5	n.a.	n.a.
New Zealand	<0.1	<0.1	0.1	n.a.	n.a.	n.a.	n.a.	n.a.	1.0	1.1	0.1	0.1	n.a.	n.a.

TABLE A1.2 (CONTINUED)

REGIONS/SUBREGIONS/COUNTRIES	NUMBER OF UNDERNOURISHED PEOPLE[1] (millions)		NUMBER OF SEVERELY FOOD-INSECURE PEOPLE[1,2] (millions)	NUMBER OF CHILDREN (UNDER 5 YEARS OF AGE) AFFECTED BY WASTING (millions)	NUMBER OF CHILDREN (UNDER 5 YEARS OF AGE) WHO ARE STUNTED (millions)		NUMBER OF CHILDREN (UNDER 5 YEARS OF AGE) WHO ARE OVERWEIGHT (millions)		NUMBER OF ADULTS (18 YEARS AND OLDER) WHO ARE OBESE (millions)		NUMBER OF WOMEN OF REPRODUCTIVE AGE (15–49) AFFECTED BY ANAEMIA (millions)		NUMBER OF INFANTS 0–5 MONTHS OF AGE EXCLUSIVELY BREASTFED (millions)	
	2004–06	2015–17	2015–17	2017[3]	2012[4]	2017[3]	2012[4]	2017[3]	2012	2016	2012	2016[5]	2012[6]	2017[7]
Oceania excluding Australia and New Zealand	n.a.	n.a.	n.a.	0.1	0.5	0.5	0.1	0.1	1.1	1.3	0.8	0.9	0.1	n.a.
Melanesia	n.a.	n.a.	n.a.	n.a.	n.a.	n.a.	n.a.	n.a.	0.9	1.1	0.7	0.9	0.1	n.a.
Fiji	<0.1	<0.1	n.a.	n.a.	n.a.	n.a.	n.a.	n.a.	0.2	0.2	0.1	0.1	n.a.	n.a.
New Caledonia	<0.1	<0.1	n.a.	n.a.	n.a.	n.a.	n.a.	n.a.	n.a.	n.a.	n.a.	n.a.	n.a.	n.a.
Papua New Guinea	n.a.	n.a.	n.a.	n.a.	0.5	n.a.	0.1	n.a.	0.7	0.9	0.6	0.7	0.1	n.a.
Solomon Islands	<0.1	<0.1	n.a.	<0.1	<0.1	<0.1	<0.1	<0.1	0.1	0.1	0.1	0.1	<0.1	<0.1
Vanuatu	<0.1	<0.1	n.a.	<0.1	<0.1	<0.1	<0.1	<0.1	<0.1	<0.1	<0.1	<0.1	<0.1	<0.1
Micronesia	n.a.	n.a.	n.a.	n.a.	n.a.	n.a.	n.a.	n.a.	0.1	0.1	<0.1	<0.1	<0.1	n.a.
Kiribati	<0.1	<0.1	n.a.	n.a.	n.a.	n.a.	n.a.	n.a.	<0.1	<0.1	<0.1	<0.1	<0.1	n.a.
Marshall Islands	n.a.	n.a.	n.a.	n.a.	n.a.	n.a.	n.a.	n.a.	<0.1	<0.1	<0.1	<0.1	<0.1	n.a.
Micronesia (Federated States of)	n.a.	n.a.	n.a.	n.a.	n.a.	n.a.	n.a.	n.a.	<0.1	<0.1	<0.1	<0.1	<0.1	n.a.
Nauru	n.a.	n.a.	n.a.	n.a.	<0.1	<0.1	<0.1	n.a.	<0.1	<0.1	<0.1	<0.1	<0.1	n.a.
Palau	n.a.	n.a.	n.a.	n.a.	n.a.	n.a.	n.a.	n.a.	<0.1	<0.1	n.a.	n.a.	n.a.	n.a.
Polynesia	<0.1	<0.1	n.a.	n.a.	n.a.	n.a.	n.a.	n.a.	0.1	0.1	<0.1	<0.1	<0.1	<0.1
American Samoa	n.a.	n.a.	n.a.	n.a.	n.a.	n.a.	n.a.	n.a.	n.a.	n.a.	n.a.	n.a.	n.a.	n.a.
Cook Islands	n.a.	<0.1	n.a.	n.a.	n.a.	n.a.	n.a.	n.a.	<0.1	<0.1	<0.1	<0.1	<0.1	n.a.
French Polynesia	<0.1	<0.1	n.a.	n.a.	n.a.	n.a.	n.a.	n.a.	n.a.	n.a.	n.a.	n.a.	n.a.	n.a.
Niue	n.a.	n.a.	n.a.	n.a.	n.a.	n.a.	n.a.	n.a.	<0.1	<0.1	<0.1	<0.1	<0.1	n.a.
Samoa	<0.1	<0.1	n.a.	<0.1	n.a.	<0.1	n.a.	<0.1	<0.1	0.1	<0.1	<0.1	<0.1	n.a.
Tokelau (Associate Member)	n.a.	n.a.	n.a.	n.a.	n.a.	n.a.	n.a.	n.a.	n.a.	n.a.	n.a.	n.a.	<0.1	n.a.
Tonga	n.a.	n.a.	n.a.	n.a.	n.a.	n.a.	n.a.	n.a.	<0.1	<0.1	<0.1	<0.1	<0.1	n.a.
Tuvalu	n.a.	n.a.	n.a.	n.a.	<0.1	<0.1	<0.1	<0.1	<0.1	<0.1	<0.1	<0.1	<0.1	n.a.

REGIONS/SUBREGIONS/COUNTRIES	NUMBER OF UNDERNOURISHED PEOPLE[1] (millions)		NUMBER OF SEVERELY FOOD-INSECURE PEOPLE[1,2] (millions)	NUMBER OF CHILDREN (UNDER 5 YEARS OF AGE) AFFECTED BY WASTING (millions)	NUMBER OF CHILDREN (UNDER 5 YEARS OF AGE) WHO ARE STUNTED (millions)		NUMBER OF CHILDREN (UNDER 5 YEARS OF AGE) WHO ARE OVERWEIGHT (millions)		NUMBER OF ADULTS (18 YEARS AND OLDER) WHO ARE OBESE (millions)		NUMBER OF WOMEN OF REPRODUCTIVE AGE (15–49) AFFECTED BY ANAEMIA (millions)		NUMBER OF INFANTS 0–5 MONTHS OF AGE EXCLUSIVELY BREASTFED (millions)	
	2004–06	2015–17	2015–17	2017[3]	2012[4]	2017[3]	2012[4]	2017[3]	2012	2016	2012	2016[5]	2012[6]	2017[7]
NORTHERN AMERICA AND EUROPE	**<26.4**	**<27.5**	**15.0**	**n.a.**	**n.a.**	**n.a.**	**n.a.**	**n.a.**	**233.1**	**255.8**	**39.4**	**44.3**	**n.a.**	**n.a.**
Northern America	**<8.2**	**<9**	**3.8**	**n.a.**	**n.a.**	**n.a.**	**n.a.**	**n.a.**	**92.0**	**102.9**	**8.7**	**10.6**	**1.1**	**1.2**
Bermuda	n.a.	n.a.	n.a.	n.a.	n.a.	n.a.	n.a.	n.a.	n.a.	n.a.	n.a.	n.a.	n.a.	n.a.
Canada	<0.8	<0.9	n.a.	n.a.	n.a.	n.a.	n.a.	n.a.	8.0	9.1	0.7	0.8	n.a.	n.a.
Greenland	n.a.	n.a.	n.a.	n.a.	n.a.	n.a.	n.a.	n.a.	n.a.	n.a.	n.a.	n.a.	n.a.	n.a.
United States of America	<7.4	<8.1	3.4	n.a.	0.4	n.a.	1.2	n.a.	84.0	93.8	8.0	9.8	1.0	1.1
Europe	**<18.3**	**<18.5**	**11.2**	**n.a.**	**n.a.**	**n.a.**	**n.a.**	**n.a.**	**141.1**	**152.9**	**30.7**	**33.7**	**n.a.**	**n.a.**
Eastern Europe	**<7.4**	**<7.3**	**3.6**	**n.a.**	**n.a.**	**n.a.**	**n.a.**	**n.a.**	**57.5**	**61.1**	**16.2**	**16.8**	**n.a.**	**n.a.**
Belarus	0.3	<0.2		n.a.	<0.1	n.a.	<0.1	n.a.	1.9	2.0	0.5	0.5	<0.1	n.a.
Bulgaria	0.5	0.2	0.2	n.a.	n.a.	n.a.	n.a.	n.a.	1.5	1.6	0.4	0.4	n.a.	n.a.
Czechia	<0.3	<0.3	<0.1	n.a.	n.a.	n.a.	n.a.	n.a.	2.3	2.5	0.6	0.6	n.a.	n.a.
Hungary	<0.3	<0.2	<0.1	n.a.	n.a.	n.a.	n.a.	n.a.	2.2	2.3	0.6	0.6	n.a.	n.a.
Poland	<1	<1	0.4	n.a.	n.a.	n.a.	n.a.	n.a.	7.3	8.0	2.2	2.4	n.a.	n.a.
Republic of Moldova	n.a.	n.a.	0.1	n.a.	<0.1	n.a.	<0.1	n.a.	0.5	0.5	0.3	0.3	<0.1	n.a.
Romania	<0.5	<0.5	0.8	n.a.	n.a.	n.a.	n.a.	n.a.	3.9	4.3	1.2	1.2	n.a.	n.a.
Russian Federation	<3.6	<3.6		n.a.	n.a.	n.a.	n.a.	n.a.	27.8	29.3	7.7	8.0	n.a.	n.a.
Slovakia	0.3	0.1	<0.1	n.a.	n.a.	n.a.	n.a.	n.a.	0.9	1.0	0.3	0.4	n.a.	n.a.
Ukraine	<1.2	1.5	1.5	n.a.	n.a.	n.a.	n.a.	n.a.	9.1	9.5	2.4	2.5	0.1	n.a.
Northern Europe	**<2.4**	**<2.6**	**2.7**	**n.a.**	**n.a.**	**n.a.**	**n.a.**	**n.a.**	**19.8**	**22.1**	**3.0**	**3.7**	**n.a.**	**n.a.**
Denmark	<0.1	<0.1	<0.1	n.a.	n.a.	n.a.	n.a.	n.a.	0.9	1.0	0.2	0.2	n.a.	n.a.
Estonia	<0.1	<0.1	<0.1	n.a.	n.a.	n.a.	n.a.	n.a.	0.2	0.2	0.1	0.1	n.a.	n.a.

TABLE A1.2 (CONTINUED)

REGIONS/SUBREGIONS/COUNTRIES	NUMBER OF UNDERNOURISHED PEOPLE[1] (millions)		NUMBER OF SEVERELY FOOD-INSECURE PEOPLE[1,2] (millions)	NUMBER OF CHILDREN (UNDER 5 YEARS OF AGE) AFFECTED BY WASTING[3] (millions)	NUMBER OF CHILDREN (UNDER 5 YEARS OF AGE) WHO ARE STUNTED (millions)		NUMBER OF CHILDREN (UNDER 5 YEARS OF AGE) WHO ARE OVERWEIGHT (millions)		NUMBER OF ADULTS (18 YEARS AND OLDER) WHO ARE OBESE (millions)		NUMBER OF WOMEN OF REPRODUCTIVE AGE (15–49) AFFECTED BY ANAEMIA[5] (millions)		NUMBER OF INFANTS 0–5 MONTHS OF AGE EXCLUSIVELY BREASTFED[6] (millions)	
	2004–06	2015–17	2015–17	2017[3]	2012[4]	2017[3]	2012[4]	2017[3]	2012	2016	2012	2016[5]	2012	2017
Finland	<0.1	<0.1		n.a.	n.a.	n.a.	n.a.	n.a.	1.0	1.1	0.2	0.2	n.a.	n.a.
Iceland	<0.1	<0.1	<0.1	n.a.	n.a.	n.a.	n.a.	n.a.	0.1	0.1	<0.1	<0.1	n.a.	n.a.
Ireland	<0.1	<0.1	0.1[d]	n.a.	n.a.	n.a.	n.a.	n.a.	0.8	1.0	0.1	0.2	n.a.	n.a.
Latvia	<0.1	<0.1	<0.1	n.a.	n.a.	n.a.	n.a.	n.a.	0.4	0.4	0.1	0.1	n.a.	n.a.
Lithuania	<0.1	<0.1	<0.1	n.a.	n.a.	n.a.	n.a.	n.a.	0.7	0.7	0.2	0.2	n.a.	n.a.
Norway	<0.1	<0.1	<0.1	n.a.	n.a.	n.a.	n.a.	n.a.	0.9	1.0	0.1	0.2	n.a.	n.a.
Sweden	<0.2	<0.2	<0.1	n.a.	n.a.	n.a.	n.a.	n.a.	1.5	1.7	0.3	0.3	n.a.	n.a.
United Kingdom	<1.5	<1.6	2.2	n.a.	n.a.	n.a.	n.a.	n.a.	13.3	15.0	1.7	2.3	n.a.	<0.1
Southern Europe	**<3.8**	**<3.8**	**2.7**	**n.a.**	**n.a.**	**n.a.**	**n.a.**	**n.a.**	**29.0**	**31.6**	**5.6**	**6.2**	**n.a.**	**n.a.**
Albania	0.3	0.2	0.3	n.a.	<0.1	n.a.	<0.1	n.a.	0.5	0.5	0.2	0.2	0.1	n.a.
Andorra	n.a.	n.a.	n.a.	n.a.	n.a.	n.a.	n.a.	n.a.	<0.1	<0.1	<0.1	<0.1	n.a.	n.a.
Bosnia and Herzegovina	0.1	<0.1	<0.1	n.a.	<0.1	n.a.	<0.1	n.a.	0.5	0.6	0.3	0.3	<0.1	n.a.
Croatia	0.1	<0.1	<0.1	n.a.	n.a.	n.a.	n.a.	n.a.	0.9	0.9	0.2	0.3	<0.1	n.a.
Greece	<0.3	<0.3	0.3	n.a.	n.a.	n.a.	n.a.	n.a.	2.3	2.5	0.3	0.4	<0.1	n.a.
Italy	<1.5	<1.5	0.6	n.a.	n.a.	n.a.	n.a.	n.a.	10.8	11.7	1.9	2.2	n.a.	n.a.
Malta	<0.1	<0.1	<0.1	<0.1	<0.1	<0.1	<0.1	n.a.	0.1	0.1	<0.1	<0.1	<0.1	n.a.
Montenegro	--	<0.1	<0.1	<0.1	<0.1	<0.1	<0.1	<0.1	0.1	0.1	<0.1	<0.1	<0.1	<0.1
Portugal	<0.3	<0.3	0.4	<0.1	<0.1	<0.1	n.a.	n.a.	1.8	2.0	0.4	0.4	n.a.	n.a.
Serbia	--	0.5	0.2	<0.1	0.1	<0.1	0.1	0.1	1.6	1.8	0.5	0.6	<0.1	<0.1
Slovenia	<0.1	<0.1	<0.1	n.a.	n.a.	n.a.	n.a.	n.a.	0.4	0.4	0.1	0.1	n.a.	n.a.
Spain	<1.1	<1.2	0.6	n.a.	n.a.	n.a.	n.a.	n.a.	9.6	10.5	1.5	1.7	n.a.	n.a.
The former Yugoslav Republic of Macedonia	0.1	<0.1	<0.1	n.a.	<0.1	n.a.	<0.1	n.a.	0.4	0.4	0.1	0.1	<0.1	n.a.

REGIONS/SUBREGIONS/COUNTRIES	NUMBER OF UNDERNOURISHED PEOPLE[1] (millions)		NUMBER OF SEVERELY FOOD-INSECURE PEOPLE[1,2] (millions)	NUMBER OF CHILDREN (UNDER 5 YEARS OF AGE) AFFECTED BY WASTING[3] (millions)	NUMBER OF CHILDREN (UNDER 5 YEARS OF AGE) WHO ARE STUNTED (millions)		NUMBER OF CHILDREN (UNDER 5 YEARS OF AGE) WHO ARE OVERWEIGHT (millions)		NUMBER OF ADULTS (18 YEARS AND OLDER) WHO ARE OBESE (millions)		NUMBER OF WOMEN OF REPRODUCTIVE AGE (15-49) AFFECTED BY ANAEMIA (millions)		NUMBER OF INFANTS 0-5 MONTHS OF AGE EXCLUSIVELY BREASTFED (millions)	
	2004-06	2015-17	2015-17	2017[3]	2012[4]	2017[3]	2012[4]	2017[3]	2012	2016	2012	2016[5]	2012[6]	2017[7]
Western Europe	**<4.6**	**<4.8**	**2.2**	**n.a.**	**n.a.**	**n.a.**	**n.a.**	**n.a.**	**34.8**	**38.1**	**6.0**	**7.0**	**n.a.**	**n.a.**
Austria	<0.2	<0.2		n.a.	n.a.	n.a.	n.a.	n.a.	1.4	1.5	0.3	0.3	n.a.	n.a.
Belgium	<0.3	<0.3		n.a.	n.a.	n.a.	n.a.	n.a.	2.0	2.2	0.3	0.4	n.a.	n.a.
France	<1.5	<1.6	0.7	n.a.	n.a.	n.a.	n.a.	n.a.	10.8	11.9	2.1	2.5	n.a.	n.a.
Germany	<2	<2	0.7	n.a.	<0.1	n.a.	0.1	n.a.	16.4	17.8	2.4	2.8	n.a.	n.a.
Luxembourg	<0.1	<0.1	<0.1	n.a.	n.a.	n.a.	n.a.	n.a.	0.1	0.1	<0.1	<0.1	n.a.	n.a.
Netherlands	<0.4	<0.4		n.a.	n.a.	n.a.	n.a.	n.a.	2.8	3.1	0.5	0.6	n.a.	n.a.
Switzerland	<0.2	<0.2	0.1	n.a.	n.a.	n.a.	n.a.	n.a.	1.3	1.4	0.3	0.4	n.a.	n.a.

[1] Regional and subregional estimates were included when more than 50 percent of population was covered. To reduce the margin of error, estimates are presented as three-year averages.

[2] FAO estimates of the number of people living in households where at least one adult has been found to be food insecure. To reduce the impact of year-to-year sampling variability, estimates are presented as three-year averages. Country-level results are presented only for those countries for which estimates are based on official national data (Ecuador, Ghana, Malawi, Republic of Korea, Saint Lucia, Seychelles and the United States of America) or as provisional estimates, based on FAO Voices of the Hungry data collected through the Gallup World Poll, for countries whose national statistical authorities (NSAs) provided permission to publish them. Note that consent to publication does not necessarily imply validation of the estimate by the NSAs and that the estimate is subject to revision as soon as suitable data from official national sources are available. Global, regional and subregional aggregates reflect data collected in approximately 150 countries.

[3] For regional estimates, values correspond to the model predicted estimate for the year 2017. For countries, the latest data available from 2013 to 2017 are used.

[4] For regional estimates, values correspond to the model predicted estimate for the year 2012. For countries, the latest data available from 2005 to 2012 are used.

[5] Anaemia data for 2016 for countries in the WHO European region are undergoing validation and thus are subject to change. The WHO European region includes: Albania, Andorra, Armenia, Austria, Azerbaijan, Belarus, Belgium, Bosnia and Herzegovina, Bulgaria, Croatia, Cyprus, Czechia, Denmark, Estonia, Finland, France, Georgia, Germany, Greece, Hungary, Iceland, Ireland, Israel, Italy, Kazakhstan, Kyrgyzstan, Latvia, Lithuania, Luxembourg, Malta, Monaco, Montenegro, Netherlands, Norway, Poland, Portugal, Republic of Moldova, Romania, Russian Federation, San Marino, Serbia, Slovakia, Slovenia, Spain, Sweden, Switzerland, Tajikistan, The former Yugoslav Republic of Macedonia, Turkey, Turkmenistan, Ukraine, United Kingdom of Great Britain and Northern Ireland, and Uzbekistan.

[6] Regional estimates are included when more than 50 percent of population is covered. For countries, the latest data available from 2005 to 2012 are used.

[7] Regional estimates are included when more than 50 percent of population is covered. For countries, the latest data available from 2013 to 2018 are used.

* Wasting, stunting and overweight under 5 years of age, and exclusive breastfeeding regional aggregates exclude Japan.

[a] The Central Agency for Public Mobilization & Statistics (CAPMAS) reports an estimate of severe food insecurity of 1.3 percent for 2015, based on HIECS data, using the WFP consolidated approach for reporting indicators of food security. Note that the two estimates are not directly comparable due to different definitions of "severe food insecurity".

[b] Consecutive low population coverage; interpret with caution.

[c] Based on a combination of official national data and FAO data.

[d] The Government of Ireland reports estimates of the "Proportion of the population at risk of food poverty" produced by the Central Statistics Office (CSO) and Economic and Social Research Institute (ESRI) as part of the Survey on Income and Social Conditions (SILC) 2015, as a proxy for SDG indicator 2.1.2. See http://irelandsdg.geohive.ie/datasets/sdg-2.1.2-prevalence-of-moderate-or-severe-food-insecurity-in-the-population-based-on-the-food-insecurity-experience-scale-nuts-3-2015-ireland-cso-amp-osi

<0.1 = less than 100 000 people.

n.a. = data not available.

METHODOLOGICAL NOTES

UNDERNOURISHMENT

Definition: Undernourishment is defined as the condition in which an individual's habitual food consumption is insufficient to provide the amount of dietary energy required to maintain a normal, active, healthy life.

How it is reported: The indicator is reported as the prevalence of undernourishment (PoU), which is an estimate of the percentage of individuals in the total population that are in a condition of undernourishment. To reduce the influence of possible estimation errors in some of the underlying parameters, national estimates are reported as a three-year moving average. Regional and global aggregates are reported as annual estimates.

Methodology: To compute an estimate of the prevalence of undernourishment in a population, the probability distribution of habitual dietary energy intake levels (expressed in Kcal per person per day) for an average individual is modelled as a parametric probability density function (pdf), *f(x)*. The indicator is obtained as the cumulative probability that daily habitual dietary energy intakes (*x*) are below minimum dietary energy requirements (MDER) (i.e. the lower bound of the acceptable range of energy requirements) for a representative average individual, as in the formula below:

$$PoU = \int_{x < MDER} f(x|\theta)dx,$$

where θ is a vector of parameters that characterizes the pdf function. In most cases, the distribution is assumed to be lognormal, and thus fully characterized by only two parameters: the mean dietary energy consumption (DEC) and its coefficient of variation (CV). In some cases, a three-parameter skew-normal or skew-lognormal distribution is considered.[351]

Data source: Different data sources can be used to estimate the different parameters of the model.

Minimum dietary energy requirements (MDER): Human energy requirements for individuals in a given sex/age class are determined on the basis of normative requirements for basic metabolic rate (BMR) per kilogram of body mass, multiplied by the ideal weight that a healthy person of that class will have, given his or her height.[352] The resulting values are multiplied by a coefficient of physical activity level (PAL) to take into account physical activity. Given that both healthy BMIs and PALs vary within groups of active and healthy individuals of the same sex and age, only a *range* of energy requirements can be computed for each sex and age group of the population. The MDER for the total population is the weighted average of the lower bounds of the energy requirement ranges for each sex and age group, with the shares of the population in each group as weights.

Information on the annual evolution in the population structure by sex and age is available for most countries in the world from the UN Department of Economic and Social Affairs (DESA) Population Prospects, produced every two years. This report uses the 2017 revision of the World Population Prospects.[353]

Information on the median height in each sex and age group is derived from a recent demographic and health survey (DHS) or other surveys that collect anthropometry data on children and adults. Even if such surveys do not refer to the same year for which the PoU is estimated, intervening changes in median heights are arguably quite small, and their impact on PoU estimates expected to be very limited.

Dietary energy consumption (DEC), coefficient of variation (CV) and skewness (Skew): When reliable data on food consumption are available from nationally representative household surveys that collect information on food consumption (for example, Living Standard Measurement Surveys or Household Incomes and Expenditure Surveys), the DEC, CV and Skew parameters can be estimated directly. However, very few countries conduct such surveys on an annual basis, leading to the need to estimate them directly or impute them for the years when no suitable survey data are available. In such cases, DEC values are estimated from the dietary energy supply (DES) reported in the Food Balance Sheets (FBS), compiled by FAO for most countries in the world (see www.fao.org/economic/ess/fbs) and available for the years up to 2016.

To impute the CV, FBS are of no use as they provide no information on the *distribution* of food consumption within a population. In the past FAO had made attempts at estimating the CV as a function of macroeconomic variables, such as per capita GDP, inequality in income (captured by the Gini index) and an index of the relative price of food.[354] The model works reasonably well to *interpolate* the values of the CV of habitual food consumption in a population for years between those when there is a survey, as the survey-based estimates can be used as anchoring points for the series of predicted CVs. However, the ability to correctly project the CV beyond the year of last available survey, with such a model, is questionable, as it would imply a high risk of meaningless out-of-sample predictions. Moreover, due to the sparsity of data on Gini indexes and to reservations about the way in which the index of the relative price of food is compiled, the benefit of using such a procedure appeared quite limited. We therefore revert to the simpler method to linearly interpolate values of the CVs in the years between surveys. The main drawback of this modelling choice is that, when the last available survey dates several years back, the value of the CV is kept constant. In such cases, changes over time in the ability to access food by different strata of the population that are not fully reflected in changes in the average national food consumption, are not reflected in PoU estimates.

PoU projections for 2017: Using the methods described above, PoU estimates are produced for all countries for which reliable FBS data are available up to 2016. To generate national-level three-year averages for 2015–17 and annual values at regional and global level in 2017, projections are needed.

As in the past editions of this report, PoU estimates for the current year are obtained by separately projecting each of the model's parameters and by applying the PoU formula presented above to the projected parameters.

Projection of the DEC. The latest available data from national food balance sheets for most countries refer to a year between 2013 and 2016. To estimate a value of DEC for up to 2017, data on the per capita availability of cereals and meats, available from the Trade and Market Division (EST) of FAO,[355] are used to estimate the likely rates of change in per capita dietary energy availability from 2013, 2014, 2015 or 2016 (depending on the country) to 2017. Such rates of change are then applied to the latest available DEC values to project them up to 2017.

Projection of the CV. As no household survey data are available for 2017, in most countries the CV estimated from last available food consumption survey data had to be projected ahead, with no change, up to 2017. However, for the countries that agreed to disseminate national estimates of their prevalence of food insecurity based on the FIES, the information could be used as auxiliary information in projecting the CV. Since 2014, FIES data provide evidence on changes in the extent of severe food insecurity that might closely reflect changes in the PoU. Such changes can be used to infer the likely changes in the CV that might have occurred in the most recent year. Recent analysis shows that, on average, CVs explain about one-third of the differences in PoU after accounting for differences in DEC and MDER. Projected changes in the CV from 2016 to 2017 for those countries are thus estimated as follows: the CV was revised by the amount that would generate a change of 1 percent in the PoU every time we observe a change of 3 percent in the prevalence of severe food insecurity (FI_{sev}).

Projection of the MDER. The MDER in 2017 is based on the projected population structure from the World Population Prospects (2017 revision, medium variant).

Challenges and limitations: While the state of being undernourished applies to individuals, due to conceptual and data-related considerations, the indicator can only refer to a population or group of individuals. The prevalence of undernourishment is thus an estimate of the percentage of individuals in a group that are in that condition – it is not based on identification of which individuals in the population are undernourished.

Due to the probabilistic nature of the inference and the margins of uncertainty associated with estimates of each of the parameters in the model, the precision of the PoU estimates is generally low. While it is not possible to calculate margins of error around PoU estimates, these would likely exceed 5 percent in most cases. For this reason, FAO does not consider national-level PoU estimates lower than 2.5 percent as sufficiently reliable to be reported.

References:

FAO. 1996. *The Sixth World Food Survey.* Rome.
L. Naiken. 2002. Keynote Paper: FAO methodology for estimating the prevalence of undernourishment. In *Measurement and Assessment of Food Deprivation and Undernutrition.* Rome, FAO.
FAO. 2014. *Refinements to the FAO Methodology for Estimating the Prevalence of Undernourishment Indicator.* FAO Statistics Division Working Paper Series. Rome.
FAO. 2014. *Advances in Hunger Measurement: Traditional FAO Methods and Recent Innovations.* FAO Statistics Division Working Paper Series. Rome.

FOOD INSECURITY AS MEASURED BY THE FOOD INSECURITY EXPERIENCE SCALE (FIES)

Definition: Food insecurity as measured by this indicator refers to limited **access to food**, at the level of individuals or households, due to lack of money or other resources. The severity of food insecurity is measured through the application of the Food Insecurity Experience Scale survey module (FIES-SM), a set of eight questions about experiences related to lack of access to food. The FIES methodology established by FAO provides a global measurement standard of food insecurity.

How it is reported: In this report, FAO provides estimates of severe food insecurity (FI_{sev}). Two estimates are reported:

▶ the **prevalence (percent) of individuals** in the population living in households where at least one adult was found to be food insecure;
▶ the **estimated number of individuals** in the population living in households where at least one adult was found to be food insecure.

Data source: Since 2014, the eight-question FIES survey module has been applied in nationally representative samples of the adult population (defined as aged 15 or older) in more than 140 countries included in the Gallup® World Poll (GWP), covering 90 percent of the world population. In most countries, samples include about 1 000 individuals, with larger samples of 3 000 individuals in India and 5 000 in mainland China.

For Ghana, Malawi (2016 and 2017), the Dominican Republic, Ecuador, Saint Lucia, Seychelles, the United States of America, (2015, 2016 and 2017) and the Republic of Korea (2014 and 2015) national government survey data were used to calculate the prevalence estimates of food insecurity by applying FAO statistical methods to adjust national results to the same global reference standard.

Methodology: The data were validated and used to construct a scale of food-insecurity severity using the Rasch model, which postulates that the probability of observing an affirmative answer by respondent i to question j is a logistic function of the distance, on an underlying scale of severity, between the position of the respondent, a_i, and that of the item, b_j.

$$Prob(X_{i,j} = \text{Yes}) = \frac{\exp(a_i - b_j)}{1 + \exp(a_i - b_j)}$$

By applying the Rasch model to the FIES data, it is possible to estimate the probability of being food insecure ($p_{i,L}$) at any given level of severity of food insecurity L, for each respondent i, with $0 < p_{i,L} < 1$.

The prevalence of food insecurity at a given level of severity (FI_L) in the population is computed as the weighted sum of the probability of being severely food insecure for all respondents (*i*) in a sample:

$$FI_L = \Sigma p_{i,L} w_i$$

where w_i are post-stratification weights that indicate the proportion of individuals or households in the national population represented by each record in the sample.

As only individuals aged 15 or more are sampled in the GWP, the prevalence estimates directly produced from these data refer to the population 15 years and older. In order to arrive at the **prevalence and number of individuals (of all ages) in the population**, an estimate is required of the number of people living in the households where at least one adult is estimated to be food insecure. This involves a multistep procedure detailed in Annex II of the Voices of the Hungry Technical Report (http://www.fao.org/3/c-i4830e.pdf).

Regional and global aggregates of food insecurity at severe levels, FI_L, are computed as:

$$FI_{L,r} = \frac{\Sigma_c FI_{L,c} \times N_c}{\Sigma_c N_c}$$

where *r* indicates the region, $FI_{L,c}$ is the value of *FI* at level *L* estimated for country *c* in the region and N_c is the corresponding population size. When no estimate of FI_L is available for a country, it is assumed to be equal to the population-weighted average of the estimated values of the remaining countries in the same region. A regional aggregate is produced only if the countries for which an estimate is available cover at least 80 percent of the region's population.

Universal thresholds are defined on the FIES global standard scale (a set of item parameter values based on results from all countries covered by the GWP in 2014–16) and converted into corresponding values on local scales. The process of calibrating each country's scale against the FIES global standard can be referred to as **equating**, and permits the production of **internationally comparable** measures of food insecurity severity for individual respondents, as well as comparable national prevalence rates.

Challenges and limitations: When food-insecurity prevalence estimates are based on FIES data collected in the GWP, with national sample sizes of about 1 000 in most countries, confidence intervals rarely exceed 20 percent of the measured prevalence (that is, prevalence rates of about 50 percent have margins of error of plus or minus 5 percent). However, confidence intervals are likely to be much smaller when national prevalence rates are estimated using larger samples and for estimates referring to subregional and regional aggregates of countries. To reduce the impact of year-to-year sampling variability, country-level estimates are presented as three-year averages.

References:
FAO. 2018. Voices of the Hungry. In: *FAO* [online]. Rome. www.fao.org/in-action/voices-of-the-hungry
FAO. 2016. *Methods for estimating comparable rates of food insecurity experienced by adults throughout the world*. Rome. http://www.fao.org/3/c-i4830e.pdf

STUNTING, WASTING AND OVERWEIGHT IN CHILDREN UNDER FIVE YEARS OF AGE

Definition of stunting: Height/length (cm) for age (months) < −2 standard deviations (SD) of the 2006 WHO Child Growth Standards median. Low height/length-for-age is an indicator that reflects the cumulative effects of undernutrition and infections since and even before birth. It may be the result of long-term nutritional deprivation, recurrent infections and lack of water and sanitation infrastructures.

How stunting is reported: The percentage of children aged 0–59 months who are below -2 SD from the median height/length-for-age of the 2006 WHO Child Growth Standards.

Definition of wasting: Weight (kg) for height/length (cm) < −2 SD of the 2006 WHO Child Growth Standards median. Low weight-for-height/

length is an indicator of acute weight loss or a failure to gain weight and can be a consequence of insufficient food intake and/or an incidence of infectious diseases, especially diarrhoea.

How wasting is reported: The percentage of children aged 0–59 months who are below -2 SD from the median weight-for-height/length of the 2006 WHO Child Growth Standards.

Definition of childhood overweight: Weight (kg) for height/length (cm) > +2 SD of the 2006 WHO Child Growth Standards median. This indicator reflects excessive weight gain for height/length generally due to energy intake exceeding children's energy requirements.

How childhood overweight is reported: The percentage of children aged 0–59 months who are over +2 SD from the median weight-for-height/length of the WHO Child Growth Standards.

Data source: UNICEF, WHO and International Bank for Reconstruction and Development/ World Bank. 2018. *UNICEF, WHO, World Bank Group Regional and Global Joint Malnutrition Estimates, May 2018 Edition* [online]. https://data. unicef.org/topic/nutrition, www.who.int/ nutgrowthdb/estimates, https://data.worldbank.org

Methodology: National nutrition surveys (MICS, DHS, national nutrition surveys, etc.) and national nutrition surveillance systems are the preferred primary data sources for child nutrition indicators. For entry in the database, they must be nationally representative, population-based surveys that present results based on the WHO Child Growth standards or provide access to the raw data enabling reanalysis.

A weighted analysis was carried out to account for the different country populations and ensure that the influence in the regional trend analysis of a country's survey estimate was proportional to the country's population. The population weights were derived from the UN Population Prospects, 2017 revision. For each data point, the respective under-five population estimate for the specific survey year was obtained. If a survey was performed over an extended period, for example

November 2013 to April 2014, the mean year in which most of the fieldwork was completed (in this case 2014) was used as the year from which to choose the respective population estimate. Weights of countries with single data points were derived by dividing the under-five population at the time of the survey by the sum of the countries' mean population in the whole region. For countries with multiple data points the weights were calculated by dividing the mean of the country's under-five population (over the observed years) by the sum of those mean populations of countries within the whole region.

A linear mixed-effect model was applied for each region or income group, using logistic transformation of prevalence and results back-transformed to original scale. The final models were then used to project the trend of malnutrition in children from 1990 to 2017. Using the resulting prevalence estimates (after back-transformation), the total numbers affected were calculated by multiplying the prevalence and lower and upper limits of the confidence intervals by the subregional population derived from the UN population estimates.

Variables: region, subregion, country, survey year, sample size, minimum and maximum age surveyed, prevalence of stunting, prevalence of wasting, prevalence of severe wasting, prevalence of overweight, country population under five years of age.

Challenges and limitations: The recommended periodicity for countries to report on stunting, overweight and wasting is every three to five years; however, for some countries data are available less frequently. While every effort has been made to maximize the comparability of statistics across countries and over time, country data may differ in terms of data collection methods, population coverage and estimation methods used. Survey estimates come with levels of uncertainty due to both sampling errors and non-sampling errors (technical measurement errors, recording errors, etc.). Neither of the two sources of error has been fully taken into account for deriving estimates at country or regional and global levels.

For the prevalence of wasting, given that surveys are generally carried out during a specific period of the year, estimates can be affected by seasonality. Seasonal factors related to wasting include food availability (e.g. preharvest periods) and disease (rainy season and diarrhoea, malaria, etc.), while natural disasters and conflicts can also show real shifts in trends that would need to be treated differently than a seasonal variation. Hence, country years' estimates for wasting might not necessarily be comparable over time. Consequently, only the most recent (2017) estimates are provided.

References:

UNICEF, WHO and World Bank. 2018. *Joint child malnutrition estimates – Levels and trends (2018 edition)* [online]. https://data.unicef.org/ topic/nutrition, www.who.int/nutgrowthdb/ estimates, https://data.worldbank.org
WHO. 2014. *Comprehensive implementation plan on maternal, infant and young child nutrition.* Geneva, Switzerland.
WHO. 2010. *Nutrition Landscape Information System (NLIS) Country Profile Indicators. Interpretation Guide.* Geneva, Switzerland.

EXCLUSIVE BREASTFEEDING

Definition: Exclusive breastfeeding for infants <6 months of age is defined as receiving only breast milk and no additional food or drink, not even water. Exclusive breastfeeding is a cornerstone of child survival and is the best food for newborns, as breast milk shapes the baby's microbiome, strengthens the immune system, and reduces the risk of developing chronic diseases.

Breastfeeding also benefits mothers by preventing post-partum haemorrhage, promoting uterine involution, decreasing risk of iron-deficiency anaemia and various types of cancer, and providing psychological benefits.

How exclusive breastfeeding is reported: Percentage of infants aged 0–5 months who are fed exclusively on breast milk with no additional food or drink – not even water – in the 24 hours preceding the survey.

Data source: UNICEF. 2018. Infant and Young Child Feeding. In: *UNICEF Data: Monitoring the Situation of Children and Women* [online]. New York. https://data.unicef.org/topic/nutrition/ infant-and-young-child-feeding

Methodology:

Infants 0–5 months of age who received only breastmilk during the previous day

---

Infants 0–5 months of age

This indicator includes breastfeeding by a wet nurse and feeding expressed breast milk.

The indicator is based on a recall of the previous day's feeding to a cross-section of infants 0–5 months of age.

In 2012, the regional and global exclusive breastfeeding estimates were generated using the most recent estimate available for each country between 2005 and 2012. Similarly, 2017 estimates were developed using the most recent estimate available for each country between 2013 and 2018. Global and regional estimates were calculated as weighted averages of the prevalence of exclusive breastfeeding in each country, using the total number of births from the World Population Prospects, 2017 revision (2012 for the baseline and 2017 for the current) as weights. Estimates are presented only where the available data are representative of at least 50 percent of corresponding regions' total number of births, unless otherwise noted.

Challenges and limitations: While a high proportion of countries collect data for exclusive breastfeeding, data are lacking in high-income countries in particular. The recommended periodicity of reporting on exclusive breastfeeding is every three to five years. However, for some countries, data are reported less frequently, meaning changes in feeding patterns are often not detected for several years after the change occurs.

Regional and global averages may be affected depending on which countries had data available for the periods considered in this report.

Using the previous day's feeding as a basis may cause the proportion of exclusively breastfed infants to be overestimated, as some infants who may have been given other liquids irregularly may not have received these in the day before the survey.

References:
UNICEF. 2018. Infant and Young Child Feeding: Exclusive breastfeeding, Predominant breastfeeding. In: *UNICEF Data: Monitoring the Situation of Children and Women* [online]. New York. https://data.unicef.org/topic/nutrition/infant-and-young-child-feeding
WHO. 2014. *Comprehensive implementation plan on maternal, infant and young child nutrition.* Geneva, Switzerland.
WHO. 2010. *Nutrition Landscape Information System (NLIS) Country Profile Indicators. Interpretation Guide.* Geneva, Switzerland.
WHO. 2008. *Indicators for assessing infant and young child feeding practices. Part 1: Definitions.* Geneva, Switzerland.

ADULT OBESITY

Definition: BMI ≥ 30.0 kg/m². The body mass index (BMI) is the ratio of weight-to-height commonly used to classify the nutritional status of adults. It is calculated as the body weight in kilograms divided by the square of the body height in meters (kg/m²). Obesity includes individuals with BMI equal to or higher than 30 kg/m².

How adult obesity is reported: Percentage of population over 18 years of age with BMI ≥ 30.0 kg/m² weighted by population.

Data source: WHO. 2017. Prevalence of obesity among adults, BMI ≥ 30, crude. In: *Global Health Observatory data repository* [online]. http://apps.who.int/gho/data/node.main.BMI30C?lang=en

Methodology: A Bayesian hierarchical model was applied to selected population-based studies that had measured height and weight in adults aged 18 years and older to estimate trends from 1975 to 2014 in mean BMI and in the prevalence of BMI categories (underweight, overweight and obesity). Overall, 1 698 population-based studies with more than 19.2 million participants aged 18 years

or older measured in 186 countries were included. The model incorporated nonlinear time trends and age patterns; national versus subnational and community representativeness; and whether data covered both rural and urban areas versus only one of them. The model also included covariates that help predict BMI, including national income, proportion of population living in urban areas, mean number of years of education, and summary measures of availability of different food types for human consumption.

Challenges and limitations: Some countries had few data sources and only 42 percent of included sources reported data for people older than 70 years.

References:
NCD Risk Factor Collaboration (NCD-RisC). 2016. Trends in adult body-mass index in 200 countries from 1975 to 2014: a pooled analysis of 1698 population-based measurement studies with 19.2 million participants. *The Lancet*, 387 (10026): 1377–1396.
WHO. 2010. *Nutrition Landscape Information System (NLIS) Country Profile Indicators. Interpretation Guide.* Geneva, Switzerland.

ANAEMIA IN WOMEN OF REPRODUCTIVE AGE

Definition: [Haemoglobin] <110g/L for pregnant women; [Haemoglobin] <120g/L for non-pregnant women. Anaemia is defined as a haemoglobin concentration below a specified cutoff point, which can change according to age, sex, physiological status, smoking habits and the altitude at which the population being assessed lives.

How anaemia is reported: Percentage of women of reproductive age (15 to 49 years old) with haemoglobin concentration below 110g/L for pregnant women and below 120 g/L for non-pregnant women.

Data sources: WHO. 2017. Prevalence of anaemia in women of reproductive age (%) (Global strategy for women's, children's and adolescents' health). In: *Global Health Observatory indicator views* [online]. http://apps.who.int/gho/data/node.imr.PREVANEMIA?lang=en

WHO. 2018. Micronutrients database. In: *Vitamin and Mineral Nutrition Information System (VMNIS)* [online]. www.who.int/vmnis/database

Methodology: National representative surveys, summary statistics from WHO's Vitamin and Mineral Nutrition Information Systems, and summary statistics reported by other national and international agencies.

Data for non-pregnant women and pregnant women were summed and weighted by the prevalence of pregnancy to generate one value for all women of reproductive age. Data were adjusted by altitude and, when available, smoking status.

Trends were modelled over time as a linear trend plus a smooth nonlinear trend, at national, regional and global levels. The model used a weighted average of various bell-shaped densities to estimate full haemoglobin distributions, which might themselves be skewed.

The estimates are also informed by covariates that help predict haemoglobin concentrations, including maternal education, proportion of population in urban areas, mean latitude, prevalence of sickle cell disorders and thalassaemia, and mean BMI. Nearly all covariates were available for every country and year, except the prevalence of sickle cell disorders and thalassaemia, which were assumed as constant over time during the analysis period for each country.

Challenges and limitations: Despite a high proportion of countries having nationally representative survey data available for anaemia, there is still a lack of reporting on this indicator, especially in high-income countries. As a result, the estimates may not capture the full variation across countries and regions, trending to "shrink" towards global means when data are sparse.

References:

G.A. Stevens, M.M. Finucane, L.M. De-Regil, C.J. Paciorek, S.R. Flaxman, F. Branca, J.P. Peña-Rosas, Z.A. Bhutta and M. Ezzati. 2013. Global, regional, and national trends in haemoglobin concentration and prevalence of total and severe anaemia in children and pregnant and non-pregnant women for 1995–2011: a systematic analysis of population-representative data. *Lancet Global Health*, 1(1): e16–25.
WHO. 2015. *The Global Prevalence of Anaemia in 2011*. Geneva, Switzerland.
WHO. 2014. *Comprehensive implementation plan on maternal, infant and young child nutrition.* Geneva, Switzerland.
WHO. 2010. *Nutrition Landscape Information System (NLIS) Country Profile Indicators. Interpretation Guide*. Geneva, Switzerland.

ANNEX 2

COUNTRY GROUP DEFINITIONS AND LISTS IN PART 2

A. Weather, climate and climate change

Weather describes conditions in the atmosphere over a short period of time (minutes to days), whereas **climate** describes the slowly varying aspects of the atmosphere–hydrosphere–land surface system and is typically characterized in terms of suitable averages of the climate system over periods of a month or more.[356] Part 2 does not analyse individual or specific weather events but instead focuses on climate variability and extremes (see below definitions) and their impact on food security and nutrition.

B. Definitions of climate variability and extremes

Climate variability refers to variations in the mean state and other statistics (standard deviations, **the occurrence of extremes**, etc.) of the climate on all spatial and temporal scales beyond that of individual weather events. Variability may be due to natural internal processes within the climate system (internal variability), or to variations in natural or anthropogenic external forcing (external variability).

Climate extremes refer to the occurrence of a value of a weather or climate variable above (or below) a threshold value near the upper (or lower) ends of the range of observed values of the variable. For simplicity, both extreme weather events and extreme climate events are referred to collectively as "climate extremes" as well as being referred to as **climate shocks**.[357]

Climate extremes analysed in Part 2 of this report – including heat spells, droughts, floods, and storms – are measured as the occurrence of any of these extremes in a country for each year of the time frame considered (1996–2016). Climate extremes are measured as the occurrence of any of these four extreme climate events, and are reported yearly for each country. Four subperiods are used: 1996–2000; 2001–2005; 2006–2010; and 2011–2016. Note that, due to data limitations, it is not possible to count the total number of climate extreme events in any given year.

Part 2 also analyses **inter-seasonal variability**, in terms of late/early start of season and the growing season length. Although such variations generally do not register as extreme weather events, they are aspects of climate variability on shorter time scales that affect the growth of crops and availability of pasture for livestock, thereby impacting on food security and nutrition. Between-season variations are defined using phenological variables derived from the vegetation index NDVI: i) a dominant reduction in the length of the seasons is defined as when a significant trend of decreased length during the period 2003–2016 involves at least 10 percent of cropland and rangeland areas of a country; ii) delay in or early onset of the seasons denotes countries where at least 10 percent of cropland and/or rangeland areas are characterized by a delayed or early onset of the season during the period 2003–2016.

C. Exposure and vulnerability to climate extremes

Whether climate variability and extremes negatively affect people's food security and nutrition depends on the **frequency and intensity of climate shocks**, the degree of **exposure** to climate shocks and their **vulnerability** to these shocks.

This analysis is undertaken on low- and middle-income countries, where there are generally – though not exclusively – higher levels of undernourishment. Out of the 140 countries classified by the World Bank as low- and middle-income, the present analysis focuses on 129 countries. Eleven countries have been omitted from the analysis since climate information is not available for them: Grenada, Maldives, Marshall Islands, Mauritius, Micronesia (Federated States of), Nauru, Saint Lucia, Saint Vincent and the Grenadines, Sao Tome and Principe, Tonga and Tuvalu. In addition, analysis shown in Figure 26 and 27 is based on 128 countries, since PoU data for Kosovo is not available.

C.1 Country exposure to climate extremes
Exposure is defined as the presence of people; livelihoods; species or ecosystems; environmental functions, services, and resources; infrastructure; or economic, social, or cultural assets in places and settings that could be adversely affected (see Annex 4 Glossary). For the purposes of the analysis in Part 2, country exposure to climate extremes is conceived as a combined measure of both the frequency and intensity of climate extremes over the areas that could be most adversely affected, as it relates most directly to impacts on food security and agricultural areas.

Exposure to heat spells is defined when the percentage of very hot days (temperature above the 90th percentile) over agriculture cropping areas is greater than 1 standard deviation (SD) in a given year/country compared to the long-term temperature average.

Exposure to drought is defined in two different ways: based on precipitation for years 1996–2005 and based on ASAP frequency of drought conditions for years 2006–2016. Exposure to drought is defined as when i) rainfall in a given

country/year over agriculture cropping areas is lower than 1 standard deviation (SD) with respect to the long-term rainfall average, or when ii) the ASAP system indicates drought conditions occuring for more than 15 percent of the growing season of croplands or rangelands in a given country/year. Although ASAP is considered to provide a more accurate measure of drought, it has only been available since 2006. Several robustness checks were performed and confirm the validity of using both ASAP and precipitation for the earlier period to identify exposure to drought.

Exposure to floods is defined as when the rainfall in a given country/year over agriculture cropping areas is greater than 2 standard deviations (SD) with respect to the long-term rainfall average in the country.

Exposure to storms is defined based on the EM-DAT datasets of medium- and large-scale disasters. Exposure to storms is defined as when in a given country/year storms have produced at least one of the following effects: i) deaths of ten or more people; ii) 100 or more people affected/injured/homeless; iii) declaration by the country of a state of emergency or an appeal for international assistance.

Countries with high exposure to climate extremes
Defined as low- and middle-income countries and territories exposed to climate extremes for at least 66 percent of the time, or more than three out of six years during the most recent subperiod of six consecutive years (2011–2016). There are 51 low- and middle-income countries that meet these criteria. For a complete list, see Table A2.2.

Countries with low exposure to climate extremes
Defined as low- and middle-income countries and territories exposed to climate extremes for up to 50 percent of the time, or less than four out of six years during the most recent subperiod of six

consecutive years (2011–2016). There are 78 low- and middle-income countries that meet these criteria.

C.2 Countries with high vulnerability to climate extremes

Vulnerability refers to the conditions that increase the probability that climate extremes will negatively affect food security (see Annex 4 Glossary). Although there are many other vulnerability factors, those below have been selected for analysis due to their relative importance for food availability and access as identified in Part 2 of the report.

Vulnerability related to climate-sensitive production and/or yields: Defined as low- and middle-income countries with at least part of their national cereal production or yield variance explained by climate factors – i.e. there is a high and statistically significant association between temperature, rainfall and vegetation growth (see Annex 3 for methodology and Table A2.1 column A for list of countries).

Vulnerability related to severe drought food security sensitivity: Countries with severe drought warnings corresponding with the occurrence of PoU change points (see Annex 3 for methodology and Table A2.1 column B for list of countries).

Vulnerability related to high dependence on agriculture: Countries with a high dependence on agriculture, with 60 percent or more people employed in the agriculture sector in 2017 – as measured by World Bank (2017) – so it is expected they are deriving their livelihood and income from the sector (see Table A2.1 column D for list of countries).

For a full description of the methodology and results, see C. Holleman, F. Rembold and O. Crespo (forthcoming). *The impact of climate variability and extremes on agriculture and food security: an analysis of the evidence and case studies*. FAO Agricultural Development Economics Technical Study 4. Rome, FAO.

TABLE A2.1
LIST OF COUNTRIES BY FOOD SECURITY VULNERABILITY FACTORS

A. Climate-sensitive production and/or yields (N = 46)	B. Severe drought food security sensitivity (N = 27)	C. Climate-sensitive production/yields and severe drought food security sensitivity (N = 16)	D. High dependence on agriculture (N = 34)
Afghanistan	Armenia	Bangladesh	Afghanistan
Algeria	Bangladesh	Belize	American Samoa
Angola	Belize	Benin	Burundi
Argentina	Benin	Cameroon	Cabo Verde
Azerbaijan	Cameroon	Central African Republic	Cameroon
Bangladesh	Central African Republic	Côte d'Ivoire	Central African Republic
Belize	Chad	Eswatini	Chad
Benin	Congo	Madagascar	Democratic People's Republic of Korea
Botswana	Côte d'Ivoire	Mauritania	Democratic Republic of the Congo
Brazil	Eritrea	Mozambique	Dominica
Burkina Faso	Gabon	Namibia	Equatorial Guinea
Cameroon	Guinea-Bissau	Panama	Eritrea
Central African Republic	Madagascar	Venezuela (Bolivarian Republic of)	Eswatini
Costa Rica	Mauritania	Yemen	Ethiopia
Côte d'Ivoire	Mozambique	Zambia	Guinea
Democratic Republic of the Congo	Namibia	Zimbabwe	Guinea-Bissau
Egypt	Nigeria		Kiribati
Eswatini	Panama		Lao People's Democratic Republic
Georgia	South Africa		Madagascar
Ghana	Togo		Malawi
Guinea	Turkmenistan		Mali
Guyana	Ukraine		Mauritania
Haiti	United Republic of Tanzania		Mozambique
Honduras	Venezuela (Bolivarian Republic of)		Nepal
Jamaica	Yemen		Niger
Lesotho	Zambia		Rwanda
Liberia	Zimbabwe		Sierra Leone
Madagascar			Solomon Islands
Malawi			Somalia
Malaysia			South Sudan
Mauritania			Uganda
Mexico			United Republic of Tanzania
Mozambique			Vanuatu
Namibia			Zimbabwe
Panama			
Paraguay			
Russian Federation			
Rwanda			
Somalia			
Suriname			
Syrian Arab Republic			
Uganda			
Venezuela (Bolivarian Republic of)			
Yemen			
Zambia			
Zimbabwe			

COUNTRIES WITH HIGH EXPOSURE TO CLIMATE EXTREMES DURING 2011–2016, BY INTER-SEASONAL VARIABILITY, FREQUENCY AND INTENSITY OF EXTREMES AND VULNERABILITY TO CLIMATE AND CONFLICT

| COUNTRIES WITH HIGH EXPOSURE TO CLIMATE EXTREMES LIST 2017 | High exposure to climate variability and extremes | | | | Vulnerability | | | | | |
| | Climate extremes | | Inter-seasonal variability | | | | | | | |
	NUMBER OF YEARS WITH EXPOSURE TO CLIMATE EXTREMES (2011–2016)	MULTIPLE TYPES OF CLIMATE EXTREMES (2011–2016)[1]	COUNTRIES WITH DELAY AND/OR EARLY START OF THE SEASON (2003–2016)	COUNTRIES WITH DECREASED LENGTH OF THE SEASON (2003–2016)	CLIMATE-SENSITIVE PRODUCTION/YIELDS (2001–2017)	CLIMATE-SENSITIVE IMPORTS (2001–2017)[2]	SEVERE DROUGHT FOOD SECURITY SENSITIVITY (2006–2015)	HIGH DEPENDENCE ON AGRICULTURE (2017)	LOW-INCOME COUNTRIES[3]	COUNTRIES AFFECTED BY CONFLICT[4]
Afghanistan	4	DSH			●	●		●	●	●
Algeria	4	DH			●					●
Bangladesh	6	S	●	●	●		●			
Belize	4	DFSH			●		●			
Bosnia and Herzegovina	4	FH								
Brazil	4	SH			●					
Bulgaria	4	DFSH								
Central African Republic	5	SH	●		●		●	●	●	●
Chad	6	DFH	●	●		●	●	●	●	●
China	6	DFSH				●				
Congo	4	DH	●				●			
Croatia	4	FH								
Cuba	5	DSH								
Democratic People's Republic of Korea	6	DFSH						●	●	
Dominican Republic	4	DSH								
Eritrea	4	DH	●	●		●	●	●	●	●
Georgia	4	DSH			●	●				●
Ghana	4	DH	●		●	●				
Guatemala	4	SH	●							
Haiti	4	DSH			●				●	
India	6	DFS								●
Indonesia	4	SH								●
Iran (Islamic Republic of)	4	DSH	●	●		●				
Kyrgyzstan	4	SH								
Lebanon	4	DFSH	●			●				
Lesotho	4	DSH	●	●	●	●				
Libya	4	DH				●				●
Madagascar	6	DSH	●	●	●	●	●	●	●	
Malawi	4	DSH	●	●		●		●	●	
Mexico	4	DFH			●	●				
Morocco	4	DSH				●				
Mozambique	4	DSH	●	●		●	●	●	●	
Myanmar	4	DFSH				●				●

TABLE A2.2
(CONTINUED)

| COUNTRIES WITH HIGH EXPOSURE TO CLIMATE EXTREMES LIST 2017 | High exposure to climate variability and extremes | | | | Vulnerability | | | | | |
| | Climate extremes | | Inter-seasonal variability | | | | | | | |
	NUMBER OF YEARS WITH EXPOSURE TO CLIMATE EXTREMES (2011–2016)	MULTIPLE TYPES OF CLIMATE EXTREMES (2011–2016)[1]	COUNTRIES WITH DELAY AND/OR EARLY START OF THE SEASON (2003–2016)	COUNTRIES WITH DECREASED LENGTH OF THE SEASON (2003–2016)	CLIMATE-SENSITIVE PRODUCTION/YIELDS (2001–2017)	CLIMATE-SENSITIVE IMPORTS (2001–2017)[2]	SEVERE DROUGHT FOOD SECURITY SENSITIVITY (2006–2015)	HIGH DEPENDENCE ON AGRICULTURE (2017)	LOW-INCOME COUNTRIES[3]	COUNTRIES AFFECTED BY CONFLICT[4]
Namibia	4	DFH			•	•	•			
Nigeria	4	DSH	•	•		•	•			•
Papua New Guinea	4	DSH								
Paraguay	4	FSH			•					
Philippines	6	FSH	•	•						•
Somalia	5	DSH	•	•	•			•	•	•
South Africa	5	DSH	•	•		•	•			
Sri Lanka	4	DFSH				•				•
Sudan	4	DSH	•	•						•
Tajikistan	4	DH								•
Thailand	4	DFSH								•
Togo	4	DH	•				•		•	
Tunisia	4	DH		•						
Turkmenistan	5	DH				•	•			
Uganda	4	DFSH			•			•	•	•
Uzbekistan	6	DH								•
Viet Nam	6	DSH			•					
Yemen	5	DSH			•		•			•
Total = 51			**19**	**14**	**19**	**22**	**14**	**10**	**12**	**21**

NOTES:
[1] D: drought; F: flood; H: heat spell; S: storm.
[2] Low- and middle-income countries with at least part of their cereal imports variance explained by climate factors, i.e. there is a statistically significant association between temperature, rainfall and vegetation growth. For methodology and results see: C. Holleman, F. Rembold and O. Crespo (forthcoming). *The impact of climate variability and extremes on agriculture and food security: an analysis of the evidence and case studies.* FAO Agricultural Development Economics Technical Study 4. Rome, FAO.
[3] Low-income countries as defined by the World Bank (https://datahelpdesk.worldbank.org/knowledgebase/articles/906519-world-bank-country-and-lending-groups).
[4] Countries affected by conflict and fragility as defined in FAO, IFAD, UNICEF, WFP and WHO. 2017. *The State of Food Security and Nutrition in the World 2017. Building resilience for peace and food security.* Rome, FAO, see Annex 2.
SOURCE: C. Holleman, F. Rembold and O. Crespo (forthcoming). *The impact of climate variability and extremes on agriculture and food security: an analysis of the evidence and case studies.* FAO Agricultural Development Economics Technical Study 4. Rome, FAO.

ANNEX 3

METHODOLOGY PART 2

A. Climate variability influence on production and imports

Data analysis was carried out to compare total cereal production and import data from FAO GIEWS Cereal Balance Sheets for the period 2001–2017 and for low- and middle-income countries, with selected weather and biophysical indicators including: annual cumulative precipitation; mean annual temperature; cumulative Normalized Difference Vegetation Index (NDVI) during active crop seasons; and drought indicators from the Anomaly Hotspots of Agriculture Production (ASAP) and Agriculture Stress Index System (ASIS). Climate data are aggregated over cropping areas smoothed for small geographical scale events, especially in large countries. NDVI is cumulative for the average crop season, while the other indicators are aggregated over the whole year.

The analysis of the climate variability influence on production and imports was performed by applying a classic correlation analysis requesting a significance of at least 90 percent. The production and import data time series have been detrended by applying a LOESS approach.[358] Detrending refers to removing a trend from a time series, where a trend usually refers to a change in the mean over time.

Countries are mapped according to their Pearson's coefficient of correlation. Figures 29 and 31 show in white the countries where part of the production (or imports) variability is explained by climate indicators (and the correlation is statistically significant). The colours denote the sign of the correlation (green = positive, red = negative).

B. PoU change point analysis data and methodology

Change points in the PoU time series were identified by applying the multiple structural changes model proposed by Bai and Perron (1998).[359] This involves finding the "best" combination of n possible breaks subject to the constraint that distance between break intervals should be above a minimum length. Here "best" means minimum sum of squared residuals from an OLS regression of PoU on a set of dummies indicating the timing of the breaks. A minimum break interval of three years was imposed in the identification of the optimal segmentation. PoU in years 2005–2016 was used to identify change points between 2006–2015. An additional constraint has been used to identify the relevant change points, i.e. only those characterized by a subsequent increasing tendency (estimated by an ordinary least squares method) have been retained.

Out of the identified PoU change points for all the low- and middle-income countries, we select those that temporally corresponded to a year ranked among the first four with the most severe frequencies of drought conditions for each country.

The frequency of drought conditions for a country is defined according to the Anomaly Hotspots of Agriculture Production (ASAP) early warning system, developed by the European Commission Joint Research Centre. ASAP drought frequency is based on the percentage of total time of the year for which a relevant share of crop or rangeland areas (> 25 percent) is affected by drought warnings according to anomalies of rainfall and NDVI.

TABLE A3.1
COUNTRIES WITH PoU CHANGE POINT CORRESPONDING TO ASAP SEVERE DROUGHT CONDITIONS

Year	Country	Group	Rank[1]	ASAP mean
2008	Armenia	Lower-middle-income	1	24.69
2010	Belize	Upper-middle-income	1	5.37
2011	Central African Republic	Low-income	1	5.21
2015	Chad	Low-income	1	22.04
2014	Mauritania	Lower-middle-income	1	26.64
2015	Mozambique	Low-income	1	28.31
2014	Panama	Upper-middle-income	1	9.90
2006	Ukraine	Lower-middle-income	1	15.58
2015	Zambia	Lower-middle-income	1	24.15
2015	Cameroon	Lower-middle-income	2	20.05
2014	Eritrea	Low-income	2	36.37
2015	Nigeria	Lower-middle-income	2	28.61
2015	Togo	Low-income	2	14.05
2015	Turkmenistan	Upper-middle-income	2	20.52
2014	Venezuela (Bolivarian Republic of)	Upper-middle-income	2	36.84
2015	Zimbabwe	Low-income	2	24.54
2007	Belize	Upper-middle-income	3	4.30
2015	Benin	Low-income	3	19.62
2015	Côte d'Ivoire	Lower-middle-income	3	9.97
2015	Madagascar	Low-income	3	17.24
2006	United Republic of Tanzania	Low-income	3	25.92
2006	Bangladesh	Lower-middle-income	4	11.56
2015	Congo	Lower-middle-income	4	6.32
2015	Gabon	Upper-middle-income	4	5.55
2012	Guinea-Bissau	Low-income	4	1.52
2006	Namibia	Upper-middle-income	4	20.33
2015	South Africa	Upper-middle-income	4	25.93
2014	Yemen	Lower-middle-income	4	10.15

NOTE:
[1] The four most severe frequencies (rank) of drought conditions for each country.
SOURCE: C. Holleman, F. Rembold and O. Crespo (forthcoming). *The impact of climate variability and extremes on agriculture and food security: an analysis of the evidence and case studies.* FAO Agricultural Development Economics Technical Study 4. Rome, FAO.

Figure 23 shows the number of countries with PoU change points that occurred in correspondence with severe drought conditions by year. The list of countries is shown in Table A3.1.

For a full description of the methodology and results, see: C. Holleman, F. Rembold and O. Crespo (forthcoming). *The impact of climate variability and extremes on agriculture and food security: an analysis of the evidence and case studies.* FAO Agricultural Development Economics Technical Study 4. Rome, FAO.

ANNEX 4

GLOSSARY

Acute food insecurity:
Food insecurity found in a specified area at a specific point in time and of a severity that threatens lives or livelihoods, or both, regardless of the causes, context or duration. Has relevance in providing strategic guidance to actions that focus on short-term objectives to prevent, mitigate or decrease severe food insecurity that threatens lives or livelihoods.[360]

Acute malnutrition:
Acute malnutrition in this report refers to extreme thinness (low weight-for-height) of individuals reflecting a reduction or loss of body weight. Child wasting, defined as weight-for-height below minus two standard deviations from the median weight-for-height in the reference population, is considered a relevant indicator of acute malnutrition, as well as small mid-upper arm circumference and bilateral pitting oedema.

Absorptive capacity:
The capacity to withstand threats and minimize exposure to shocks and stressors through preventative measures and appropriate coping strategies to avoid permanent, negative impacts.[361] The capacity to absorb shocks and stresses by increasing access to climate risk insurance and social protection systems.[362]

Adaptation:
The process of adjustment to actual or expected climate and its effects. In human systems, adaptation seeks to moderate or avoid harm or exploit beneficial opportunities. In some natural systems, human intervention may facilitate adjustment to expected climate and its effects.[363]

Adaptive capacity:
The ability of systems, institutions, humans and other organisms to adjust to potential damage, to take advantage of opportunities, or to respond to consequences.[364] The ability of a system to adjust to climate change (including climate variability and extremes) in order to moderate potential damages, to take advantage of opportunities, or to cope with the consequences.[365] The capacity to adapt to new options in the face of crisis by making proactive and informed choices about alternative livelihood strategies based on an understanding of changing conditions.[366]

Anomaly:
The difference between a climate variable averaged over a particular period (e.g. for a particular year or group of years), and the same climate variable averaged over a longer (baseline/reference) period (e.g. averaged over the 35 years between 1981 and 2016).

Anthropogenic:
Resulting from or produced by human activities.[367]

Anthropometry:
Use of human body measurements to obtain information about nutritional status.

Capacity:
The combination of all the strengths, attributes and resources available within an organization, community or society to manage and reduce disaster risks and strengthen resilience. Capacity may include infrastructure, institutions, human knowledge and skills, and collective attributes such as social relationships, leadership and management.[368]

Chronic food insecurity:
Food insecurity that persists over time mainly due to structural causes. Can include seasonal food insecurity found in periods with non-exceptional conditions. Has relevance in providing strategic guidance to actions that focus on the medium- and long-term improvement of the quality and quantity of food consumption for an active and healthy life.[369]

Climate:
Climate in a narrow sense is usually defined as the average weather, or more rigorously, as the statistical description in terms of the mean and variability of relevant quantities over a period of time ranging from months to thousands or millions of years.[370]

Climate change:
Climate change refers to a change in the state of the climate that can be identified (e.g. by using statistical tests) by changes in the mean and/or the variability of its properties, and that persists for an extended period, typically decades or longer.[371]

Climate change adaptation (CCA):
An approach to adaptation (see adaptation definition above) that addresses current or expected climate variability and changing average climate conditions.

Climate extreme (extreme weather or climate event):
The occurrence of a value of a weather or climate variable above (or below) a threshold value near the upper (or lower) ends of the range of observed values of the variable. For simplicity, both extreme weather events and extreme climate events are referred to collectively as "climate extremes".[372]

Climate resilience:
An approach to building and/or strengthening resilience (see resilience definition below) that addresses current or expected climate variability and changing average climate conditions.

Climate-resilient pathways:
Iterative processes for managing change within complex systems in order to reduce disruptions and enhance opportunities associated with climate change.[373]

Climate services:
Climate services involve the production, translation, transfer and use of climate knowledge and information to support decision-making by individuals and organizations. Information needs to be easily accessible, timely, easy to understand and relevant to users so they can use it to take action.

Climate variability:
Refers to variations in the mean state and other statistics (standard deviations, the occurrence of extremes, etc.) of the climate on all spatial and temporal scales beyond that of individual weather events. Variability may be due to natural internal processes within the climate system (internal variability), or to variations in natural or anthropogenic external forcing (external variability).[374]

Climate shocks:
Climate shocks include not only those disturbances in the usual pattern of rainfall and temperatures but also complex events like droughts and floods. Equivalent to the concept of a natural hazard or stress, an exogenous event that can have a negative impact on food and nutrition security, depending on the vulnerability of an individual, a household, a community, or systems to the shock.[375]

Climatology:
The average of a climate-related variable over a long period of time, e.g. averaged over 30 years between 1981 and 2010.

Coping capacity:
The ability of people, institutions, organizations and systems, using available skills, values, beliefs, resources and opportunities, to address, manage and overcome adverse conditions in the short to medium term.[376]

Dietary energy intake:
The energy content of food consumed.

Dietary energy supply (DES):
Food available for human consumption, expressed in kilocalories per person per day (kcal/person/day). At the country level, it is calculated as the food remaining for human use after deduction of all non-food utilizations (i.e. food = production + imports + stock withdrawals – exports – industrial use – animal feed – seed – wastage – additions to stock). Wastage includes loss of usable products occurring along distribution chains from farm gate (or port of import) up to retail level.

Disaster risk management (DRM):
Disaster risk management is the application of disaster risk reduction policies and strategies to prevent new disaster risk, reduce existing disaster risk and manage residual risk, contributing to the strengthening of resilience and reduction of disaster losses.[377]

Disaster risk reduction (DRR):
Disaster risk reduction is aimed at preventing new and reducing existing disaster risk and managing residual risk, all of which contribute to strengthening resilience and therefore to the achievement of sustainable development. Disaster risk reduction is the policy objective of disaster risk management, and its goals and objectives are defined in disaster risk reduction strategies and plans.[378]

Drought:
A period of abnormally dry weather long enough to cause a serious hydrological imbalance. A period with an abnormal precipitation deficit is defined as a meteorological drought.[379]

Early warning system (EWS):
The set of capacities needed to generate and disseminate timely and meaningful warning information so that individuals, communities and organizations threatened by a hazard can prepare prompt and appropriate action to reduce the possibility of harm or loss.[380]

El Niño–Southern Oscillation (ENSO):
The term El Niño is used to describe a basin-wide warming of the tropical Pacific Ocean east of the International Date Line. This oceanic event is associated with a fluctuation of a global-scale tropical and subtropical surface pressure pattern called the Southern Oscillation. This combined atmospheric–oceanic phenomenon, usually occurring around every two to seven years, is known as the El Niño-Southern Oscillation (ENSO). The cold phase of ENSO is called La Niña.[381]

Exposure:
The presence of people, livelihoods, species or ecosystems, environmental functions, services and resources, infrastructure, or economic, social or cultural assets in places and settings that could be adversely affected.[382]

Extreme weather or climate event:
The occurrence of a value of a weather or climate variable above (or below) a threshold value near the upper (or lower) ends of the range of observed values of the variable. Many weather and climate extremes are the result of natural **climate variability** (including phenomena such as ENSO), and natural decadal or multi-decadal variations in the climate provide the backdrop for anthropogenic **climate changes**. Even if there were no anthropogenic changes in climate, a wide variety of natural weather and climate extremes would still occur.

Flood:
The overflowing of the normal confines of a stream or other body of water, or the accumulation of water over areas not normally submerged. Floods include river (fluvial) floods,

flash floods, urban floods, pluvial floods, sewer floods, coastal floods, and glacial lake outburst floods.[383]

Food insecurity:
A situation that exists when people lack secure access to sufficient amounts of safe and nutritious food for normal growth and development and an active and healthy life. It may be caused by unavailability of food, insufficient purchasing power, inappropriate distribution or inadequate use of food at the household level. Food insecurity, poor conditions of health and sanitation and inappropriate care and feeding practices are the major causes of poor nutritional status. Food insecurity may be chronic, seasonal or transitory.

Food security:
A situation that exists when all people, at all times, have physical, social and economic access to sufficient, safe and nutritious food that meets their dietary needs and food preferences for an active and healthy life. Based on this definition, four food security dimensions can be identified: food availability, economic and physical access to food, food utilization and stability over time.

Food security dimensions:
Refers to the four dimensions of food security:
▶ **Availability** – This dimension addresses whether or not food is actually or potentially physically present, including aspects of production, food reserves, markets and transportation, and wild foods.
▶ **Access** – If food is actually or potentially physically present, the next question is whether or not households and individuals have sufficient access to that food.
▶ **Utilization** – If food is available and households have adequate access to it, the next question is whether or not households are maximizing the consumption of adequate nutrition and energy. Sufficient energy and nutrient intake by individuals is the result of good care and feeding practices, food preparation, dietary diversity and intra-household distribution of food. Combined with good biological utilization of food consumed, this determines the *nutritional status* of individuals.

▶ **Stability** – If the dimensions of availability, access and utilization are sufficiently met, stability is the condition in which the whole system is stable, thus ensuring that households are food secure at all times. Stability issues can refer to short-term instability (which can lead to *acute food insecurity*) or medium- to long-term instability (which can lead to *chronic food insecurity*). Climatic, economic, social and political factors can all be a source of instability.

Hazard:
A process, phenomenon or human activity that may cause loss of life, injury or other health impacts, property damage, social and economic disruption or environmental degradation.[384] Natural hazard is synonymous with "climate shock" in this report.

Heatwave:
A period of abnormally and uncomfortably hot weather.[385]

Hunger:
Hunger is an uncomfortable or painful physical sensation caused by insufficient consumption of dietary energy. In this report, the term hunger is synonymous with chronic undernourishment.

Kilocalorie (kcal):
A unit of measurement of energy. One kilocalorie equals 1 000 calories. In the International System of Units (SI), the universal unit of energy is the joule (J). One kilocalorie = 4.184 kilojoules (kJ).

Livelihood assets or capital:
The resources used and the activities undertaken in order to live. These assets are referred to as livelihood assets and in the Sustainable Livelihoods Framework are defined under the following five categories of "capital":
▶ **Economic or financial capital:** capital base (regular inflows of money, credit/debt, savings and other economic assets)
▶ **Human capital:** skills, knowledge, labour (includes good health and physical capability)
▶ **Physical capital:** productive assets, infrastructure (buildings, roads, production equipment and technologies)

▶ **Natural capital:** Natural resource stocks (land, soil, water, air, genetic resources, forests, etc.) and environmental services (hydrological cycle, pollution sinks, etc.)

▶ **Social capital:** resources (networks, social claims, social relations, affiliations, associations)

The ways in which people utilize and combine their livelihood assets to obtain food, income and other goods and services are defined as their livelihood strategies.

Macronutrients:
These are the proteins, carbohydrates and fats available to be used for energy; measured in grams.

Malnutrition:
An abnormal physiological condition caused by inadequate, unbalanced or excessive consumption of macronutrients and/or micronutrients. Malnutrition includes undernutrition and overnutrition as well as micronutrient deficiencies.

Micronutrients:
Vitamins, minerals and other substances that are required by the body in small amounts; measured in milligrams or micrograms.

Mitigation (of climate change):
A human intervention to reduce the sources or enhance the sinks of greenhouse gases that lead to climate change.[386]

Mitigation (of disaster risk and disaster):
The lessening of the potential adverse impacts of physical hazards (including those that are human-induced) through actions that reduce hazard, exposure and vulnerability.[387]

Nutrition security:
A situation that exists when secure access to an appropriately nutritious diet is coupled with a sanitary environment and adequate health services and care, in order to ensure a healthy and active life for all household members. Nutrition security differs from food security in that it also considers the aspects of adequate caregiving practices, health and hygiene in addition to dietary adequacy.

Nutrition-sensitive intervention:
An action designed to address the underlying determinants of nutrition (which include household food security, care for mothers and children and primary health care and sanitation) but not necessarily having nutrition as the predominant goal.

Nutritional status:
The physiological state of an individual that results from the relationship between nutrient intake and requirements and the body's ability to digest, absorb and use these nutrients.

Overnutrition:
A result of excessive food intake relative to dietary nutrient requirements.

Overweight and obesity:
Body weight that is above normal for height as a result of an excessive accumulation of fat. It is usually a manifestation of expending fewer calories than are consumed. In adults, overweight is defined as a BMI of more than 25 kg/m^2 but less than 30 kg/m^2, and obesity as a BMI of 30 kg/m^2 or more. In children under five years of age, overweight is defined as weight-for-height greater than 2 standard deviations above the WHO Child Growth Standards median, and obesity as weight-for-height greater than 3 standard deviations above the WHO Child Growth Standards median.

Preparedness:
The knowledge and capacities developed by governments, response and recovery organizations, communities and individuals to effectively anticipate, respond to and recover from the impacts of likely, imminent or current disasters.[388]

Prevention:
Activities and measures to avoid existing and new disaster risks. Prevention (i.e. disaster prevention) expresses the concept and intention to completely avoid potential adverse impacts of hazardous events.[389]

Resilience:
Resilience is the ability of individuals, households, communities, cities, institutions, systems and societies to prevent, resist, absorb,

adapt, respond and recover positively, efficiently and effectively when faced with a wide range of risks, while maintaining an acceptable level of functioning and without compromising long-term prospects for sustainable development, peace and security, human rights and well-being for all.[390]

Risk:
The probability or likelihood of occurrence of hazardous events or trends multiplied by the impacts if these events or trends occur. Risk to food insecurity is the probability of food insecurity resulting from interactions between a natural or human-induced hazard/shock/stress and vulnerable conditions.

Severe food insecurity:
Based on the Food Insecurity Experience Scale (FIES), someone experiencing severe food insecurity is likely to have gone entire days without eating due to lack of money or other resources (see Methodological notes, Annex 1).

Stunting:
Low height-for-age, reflecting a past episode or episodes of sustained undernutrition. In children under five years of age, stunting is defined as height-for-age less than −2 standard deviations below the WHO Child Growth Standards median.

Transformative capacity:
The capacity to transform the set of livelihood choices available through empowerment and growth, including governance mechanisms, policies/regulations, infrastructure, community networks, and formal and informal social protection mechanisms that constitute an enabling environment for systemic change.[391]

Undernourishment:
Undernourishment is defined as the condition in which an individual's habitual food consumption is insufficient to provide the amount of dietary energy required to maintain a normal, active, healthy life. For the purposes of this report, hunger is defined as being synonymous with chronic undernourishment.

Undernutrition:
The outcome of poor nutritional intake in terms of quantity and/or quality, and/or poor absorption and/or poor biological use of nutrients consumed as a result of repeated instances of disease. It includes being underweight for one's age, too short for one's age (stunted), dangerously thin for one's height (suffering from wasting) and deficient in vitamins and minerals (micronutrient deficiency).

Vulnerability:
The conditions determined by physical, social, economic and environmental factors or processes that increase the susceptibility of an individual, a community, assets or systems to the impacts of hazards.[392]

Vulnerability to food insecurity is the range of conditions that increases the susceptibility of a household to the impact on food security in case of a shock or hazard.

Wasting:
Low weight-for-height, generally the result of weight loss associated with a recent period of inadequate calorie intake and/or disease. In children under five years of age, wasting is defined as weight-for-height less than −2 standard deviations below the WHO Child Growth Standards median.

Weather:
Weather describes conditions of the atmosphere over a short period of time (minutes to days), whereas climate is how the atmosphere behaves over relatively longer periods of time (the long-term average of weather over time). The difference between weather and climate is a measure of time (see above definitions for climate, climate change, climate variability, and climate extremes).[393]

NOTES

1 Each edition of *The State of Food Security and Nutrition in the World* presents a complete revised series of the PoU indicators, as the result of the updating of all supporting evidence (see Box 1 for details). For this reason, readers are advised to always consider the PoU estimates in the most recent report and avoid comparisons with those presented in past editions.

2 United Nations Office for the Coordination of Humanitarian Affairs (OCHA). 2016. El Niño: Southern Africa faces its worst drought in 35 years. [online]. New York, USA. www.unocha.org/story/el-niño-southern-africa-faces-its-worst-drought-35-years

3 See, for example: Statistics South Africa. 2016. Consumer Price Index March 2016 [online] www.statssa.gov.za/publications/P0141/P0141March2016.pdf, Table C, page 5, showing how food has been, by far, the major contributor to the increase in the Consumer Price Index in South Africa in 2015 and 2016. See also: http://www.rbz.co.zw/assets/quarterly-economic-review-december-2017.pdf, page 21, for similar evidence of food inflation in Zimbabwe; and https://www.knbs.or.ke/download/cpi-rates-inflation-september-2017 in Kenya, for 2017.

4 See, for example, C.F. Ndife. 2017. A Comparative Study of Economic Growth in the West African States. *Journal of World Economic Research*, 6(6): 75–79.

5 United Nations (UN). 2017. *World Population Prospects 2017* [online]. New York, USA. https://esa.un.org/unpd/wpp

6 FAO. 2018. Voices of the Hungry. In: *FAO* [online]. Rome. www.fao.org/in-action/voices-of-the-hungry

7 See methodological note in Annex 1.

8 Ecuador, Ghana, Malawi, Republic of Korea, Saint Lucia, Seychelles and the United States of America.

9 See C. Cafiero, S. Viviani and M. Nord. 2018. Food security measurement in a global context: The Food Insecurity Experience Scale. *Measurement*, 116: 146–152.

10 For the countries highlighted with a red marker, the absolute difference between the estimated PoU and Fi_{sev} is larger than their average.

11 For China, the estimate of the PoU is based on official but old data on the distribution of food access in the population and thus may not reflect the considerable increase in access to food by the poor that has occurred over the last two decades. Efforts are currently in progress with the country to have access to updated food consumption data.

12 WHO and UNICEF. The extension of the 2025 Maternal, Infant and Young Child nutrition targets to 2030. Discussion Paper [online]. www.who.int/nutrition/global-target-2025/discussion-paper-extension-targets-2030.pdf

13 United Nations, General Assembly (UNGA). 2018. Implementation of the United Nations Decade of Action on Nutrition (2016–2025). A/72/829 (11 April 2018).

14 UNGA, 2018 (see note 13).

15 World Health Organization (WHO). 2018. Child growth standards. In: *The World Health Organization* [online]. Geneva, Switzerland. www.who.int/childgrowth

16 R.E. Black, C.G. Victora, S.P. Walker, Z.A. Bhutta, P. Christian, M. de Onis, M. Ezzati, S. Grantham-McGregor, J. Katz, R. Martorell, R. Uauy and Maternal and Child Nutrition Study Group. 2013. Maternal and child undernutrition and overweight in low-income and middle-income countries. *The Lancet*, 382(9890): 427–451.

17 T. Khara and C. Dolan. 2014. *Technical briefing paper: Associations between wasting and stunting, policy, programming and research implications.* Oxford, UK, Emergency Nutrition Network.

18 World Health Organization (WHO), United Nations High Commissioner for Refugees (UNHCR), International Federation of Red Cross (IFRC) and World Food Programme (WFP). 2000. *The management of nutrition in major emergencies.* Geneva, Switzerland, WHO.

19 UNICEF. 2018. *Annual Results Report 2017 – Nutrition.* New York, USA.

20 Z.A. Bhutta, J.K. Das, A. Rizvi, M.F. Gaffey, N. Walker, S. Horton, P. Webb, A. Lartey and R.E. Black. 2013. Evidence-based interventions for improvement of maternal and child nutrition: what can be done and at what cost? *The Lancet*, 382(9890): 452–477.

21 WHO. 1995. Physical status: the use and interpretation of anthropometry. Report of a WHO Expert Committee. *World Health Organ Technical Reports Series*, 854: 1–452.

22 M.C.H. Jukes, L.J. Drake and D.A.P. Bundy. 2007. School health, nutrition and education for all: levelling the playing field. *Bulletin of the World Health Organization*, 87(1): 75.

23 United Nations Children's Fund (UNICEF). 1990. *Strategy for improved nutrition of children and women in developing countries*. New York, USA.

24 High Level Panel of Experts (HLPE). 2017. *Nutrition and food systems*. A report by the High Level Panel of Experts on Food Security and Nutrition of the Committee on World Food Security. Rome.

25 B.M. Popkin, L.S. Adair and S.W. Ng. 2012. Global nutrition transition and the pandemic of obesity in developing countries. *Nutrition reviews*, 70(1): 3–21.

26 WHO. 2017. *The double burden of malnutrition*. Geneva, Switzerland.

27 GBD 2015 Mortality and Causes of Death Collaborators. 2016. Global, regional, and national life expectancy, all-cause mortality, and cause-specific mortality for 249 causes of death, 1980-2015: a systematic analysis for the Global Burden of Disease Study 2015. *The Lancet*, 388(10053): 1459–1544.

28 WHO and UNICEF. 2017. *Report of the Fourth Meeting of the WHO-UNICEF Technical Expert Advisory group on nutrition Monitoring (TEAM)*. Geneva, Switzerland.

29 WHO. 2011. *Haemoglobin concentrations for the diagnosis of anaemia and assessment of severity*. Geneva, Switzerland.

30 R. Pérez-Escamilla, O. Bermudez, G.S. Buccini, S. Kumanyika, C.K. Lutter, P. Monsivais and C. Victora. 2018. Nutrition disparities and the global burden of malnutrition. *British Medical Journal*, 361: k2252.

31 H. Ghattas. 2014. *Food security and nutrition in the context of the nutrition transition*. Technical Paper. Rome, FAO; C. Maitra. 2018. *A review of studies examining the link between food insecurity and malnutrition*. Technical Paper. Rome, FAO.

32 The FIES, FAO's global food security measure, was first introduced in 2014 and was not used in any of the studies cited. The studies cited used other experience-based food security metrics that have been used for national monitoring and research for many years. These include the US Household Food Security Survey Module, the Latin American and Caribbean Food Security Scale (*Escala Latinoamericana y Caribeña de Seguridad Alimentaria* – ELCSA), the Brazilian Food Insecurity Scale (*Escala Brasileira de Insegurança Alimentar* – EBIA) and the Household Food Insecurity Access Scale (HFIAS). These survey modules all share the same origins, are based on the same underlying concept of food insecurity and are composed of nearly identical sets of questions.

33 A review of existing evidence linking the experience of food insecurity and selected indicators of malnutrition was conducted covering a wide range of countries from almost all regions and income levels (Maitra, 2018 [see note 31]; see also Table 6).

34 Maitra, 2018 (see note 31).

35 Maitra, 2018 (see note 31); see Table 6.

36 Maitra, 2018 (see note 31).

37 Maitra, 2018 (see note 31).

38 Maitra, 2018 (see note 31).

39 Maitra, 2018 (see note 31).

40 Maitra, 2018 (see note 31).

41 B.M. Popkin, L.S. Adair and S.W. Ng. 2012. (see note 25).

42 Ghattas, 2014 (see note 31).

43 Maitra, 2018 (see note 31).

44 Maitra, 2018 (see note 31).

45 B.E. Levin. 2006. Metabolic imprinting: critical impact of the perinatal environment on the regulation of energy homeostasis. *Philosophical Transactions of the Royal Society B: Biological Sciences*, 361(1471): 1107–1121; R. Pérez-Escamilla, O. Bermudez, G.S. Buccini, S. Kumanyika, C.K. Lutter, P. Monsivais and C. Victora. 2018. Nutrition disparities and the global burden of malnutrition. *British Medical Journal*, 361: k2252.

NOTES

46 Maitra, 2018 (see note 31).

47 WHO. 2016. *Report of the Commission on Ending Childhood Obesity.* Geneva, Switzerland.

48 Maitra, 2018 (see note 31).

49 WHO, 2016 (see note 47).

50 A.L. Pereira, S. Handa and G. Holmqvist. 2017. *Prevalence and correlates of food insecurity among children across the globe.* Innocenti Working Paper 2017-09. Florence, Italy, UNICEF Office of Research.

51 WHO. 2017. *Double-duty actions for nutrition. Policy brief.* Geneva, Switzerland.

52 Pérez-Escamilla *et al.*, 2018. (see note 30).

53 HLPE, 2017 (see note 24).

54 B.M. Popkin and T. Reardon. 2018. Obesity and the food system transformation in Latin America. *Obesity Reviews.* April.

55 FAO. 2016. *Climate change and food security: risks and responses.* Rome.

56 See Annex 2 for definitions of weather, climate change, climate variability, climate extremes and other climate-related terms.

57 UN. 2016. *World Economic and Social Survey 2016 – Climate Change Resilience: An Opportunity for Reducing Inequalities.* New York, USA.

58 Intergovernmental Panel on Climate Change (IPCC). 2014. *Climate Change 2014: Synthesis Report. Contribution of Working Groups I, II and III to the Fifth Assessment Report of the Intergovernmental Panel on Climate Change.* [Core Writing Team, R.K. Pachauri and L.A. Meyer, eds]. Geneva, Switzerland.

59 Centre for Research on the Epidemiology of Disasters (CRED). 2015. *The Human Cost of Natural Disaster 2015: A Global Perspective.* Brussels.

60 FAO. 2015. *The impact of disasters on agriculture and food security.* Rome.

61 Integrated Food Security Phase Classification (IPC) Phase 3 and above or equivalent.

62 Food Security Information Systems (FSIN). 2018. *Global Report on Food Crises 2018.* Rome.

63 United Nations Department of Economic and Social Affairs (UN DESA). 2017. Sustainable Development Goal 13: Take urgent action to combat climate change and its impact. In: *Sustainable Development Knowledge Platform* [online]. New York, USA. https://sustainabledevelopment.un.org/sdg13

64 IPCC, 2014 (see note 58).

65 This refers to the "pre-industrial period" as noted by the Paris agreement and 1.5 and 2 °C framings.

66 IPCC. 2014. *Climate Change 2014: Impacts, Adaptation, and Vulnerability. Part B: Regional Aspects. Contribution of Working Group II to the Fifth Assessment Report of the Intergovernmental Panel on Climate Change.* [V.R. Barros, C.B. Field, D.J. Dokken, M.D. Mastrandrea, K.J. Mach, T.E. Bilir, M. Chatterjee, K.L. Ebi, Y.O. Estrada, R.C. Genova, B. Girma, E.S. Kissel, A.N. Levy, S. MacCracken, P.R. Mastrandrea and L.L. White, eds]. Cambridge, UK, and New York, USA, Cambridge University Press.

67 IPCC. 2013. *Climate Change 2013: The Physical Science Basis. Contribution of Working Group I to the Fifth Assessment Report of the Intergovernmental Panel on Climate Change.* Cambridge, UK and New York, USA, Cambridge University Press.

68 C. Holleman, F. Rembold and O. Crespo (forthcoming). *The impact of climate variability and extremes on agriculture and food security: an analysis of the evidence and case studies.* FAO Agricultural Development Economics Technical Study 4. Rome, FAO.

69 Holleman, Rembold and Crespo (forthcoming). (see note 68).

70 S.N.A. Codjoe and G. Owusu. 2011. Climate change/variability and food systems: evidence from the Afram Plains, Ghana. *Regional Environmental Change,* 11(4): 753–765.

71 S. Adjei-Nsiah, P. Mapfumo, J.O. Fening, V. Anchirina, R.N. Issaka and K. Giller. 2010. Farmers' Perceptions of Climate Change and Variability and Existing Opportunities for Adaptation in Wenchi Area of Ghana. *The International*

Journal of Climate Change: Impacts and Responses, 2: 49–60.

72 S.L.M. Traerup and O. Mertz. 2011. Rainfall variability and household coping strategies in northern Tanzania: a motivation for district-level strategies. *Regional Environmental Change*, 11(3): 471–481; J. Tambo and T. Abdoulaye. 2013. Smallholder farmers' perceptions of and adaptations to climate change in Nigerian savanna. *Regional Environmental Change*, 11(2): 375–388.

73 Exceptions are N. Debela *et al.* (2015), where in the two decades of the study in Borno, Ethiopia (1992–2012) the rainfall was less, there were fewer rainy days and higher temperatures, in comparison to the preceding decade 1980–1992. See N. Debela, C. Mohammed, K. Bridle, R. Corkrey and D. McNeil. 2015. Perception of climate change and its impact by smallholders in pastoral/agropastoral systems of Borana, South Ethiopia. *SpringerPlus*, 4(236).

74 C. Neely, S. Bunning and A. Wilkes, eds. 2009. *Review of evidence on drylands pastoral systems and climate change: Implications and opportunities for mitigation and adaptation*. Rome, FAO.

75 United Nations Economic and Social Council (ECOSOC). 2007. *Africa Review Report on Drought and Desertification*. E/ECA/ACSD/5/3 (November 2007).

76 D. Griffin and K.J. Anchukaitis. 2014. How unusual is the 2012-2014 California drought? *Geophysical Research Letters*, 41(24): 9017–9023; WMO. 2016. Hotter, drier, wetter. Face the future [online]. www.wmo.int/worldmetday/content/hotter-drier-wetter-face-future; J. Blunden and D.S. Arndt. 2016. State of the Climate in 2015. *Bulletin of the American Meteorological Society*, 97(8): S1-S275.

77 The indicator for drought conditions frequency is extracted from Anomaly Hotspots of Agriculture Production (ASAP), an early warning system developed by the European Commission Joint Research Centre.

78 Holleman, Rembold and Crespo (forthcoming). (see note 68).

79 M. Boko, I. Niang, A. Nyong, C. Vogel, A. Githeko, M. Medany, B. Osman-Elasha, R. Tabo and P. Yanda. 2007. Africa. In IPCC. 2007. *Climate Change 2007: impacts, adaptation and vulnerability. Contribution of Working Group II to the Fourth Assessment Report of the Intergovernmental Panel on Climate Change*, pp. 433–467 [M.L. Parry,

O.F. Canziani, J.P. Palutikof, P.J. van der Linden and C.E. Hanson, eds]. Cambridge, UK, Cambridge University Press.

80 J. Syvitski, A. Kettner, I. Overeem, E. Hutton, M. Hannon, R. Brakenridge, J. Day *et al.* 2009. Sinking Deltas due to Human Activities. *Nature Geoscience*, 2(10).

81 A. Revi, D.E. Satterthwaite, F. Aragón-Durand, J. Corfee-Morlot, R.B.R. Kiunsi, M. Pelling, D.C. Roberts and W. Solecki. 2014. Urban areas. In IPCC. 2014. *Climate Change 2014: Impacts, Adaptation, and Vulnerability. Part A: Global and Sectoral Aspects. Contribution of Working Group II to the Fifth Assessment Report of the Intergovernmental Panel on Climate Change*, pp. 535–612 [C.B. Field, V.R. Barros, D.J. Dokken, K.J. Mach, M.D. Mastrandrea, T.E. Bilir, M. Chatterjee, K.L. Ebi, Y.O. Estrada, R.C. Genova, B. Girma, E.S. Kissel, A.N. Levy, S. MacCracken, P.R. Mastrandrea and L.L. White, eds]. Cambridge, UK and New York, USA, Cambridge University Press.

82 P.P. Wong, I.J. Losada, J.-P. Gattuso, J. Hinkel, A. Khattabi, K.L. McInnes, Y. Saito and A. Sallenger. 2014. Coastal systems and low-lying areas. In IPCC. 2014. *Climate Change 2014: Impacts, Adaptation, and Vulnerability. Part A: Global and Sectoral Aspects. Contribution of Working Group II to the Fifth Assessment Report of the Intergovernmental Panel on Climate Change*, pp. 361–409 [Field, C.B., V.R. Barros, D.J. Dokken, K.J. Mach, M.D. Mastrandrea, T.E. Bilir, M. Chatterjee, K.L. Ebi, Y.O. Estrada, R.C. Genova, B. Girma, E.S. Kissel, A.N. Levy, S. MacCracken, P.R. Mastrandrea and L.L. White, eds]. Cambridge, UK and New York, USA, Cambridge University Press.

83 Precisely because undernourishment is estimated at the national scale, it requires significant changes in food availability or access to affect national-level estimates, and changes usually occur over time.

84 The PoU estimates the proportion of the population habitually meeting the (average) minimum daily dietary intake requirements. It uses the DEC (mean dietary energy consumption), which is computed as a three-year average. This means that the PoU is a highly smoothed data time series, which can be expected to reflect to some extent major variations in production, in those cases where a country is not able to compensate large production drops with stocks and imports. Although the way the PoU is computed and smoothed over a three-year period makes direct regression with climate indicators inappropriate, it is possible to examine whether major climate shocks, such as extreme droughts, can be linked to change points in PoU.

85 Severe drought years are defined here as those belonging to the first four ranks of ASAP drought conditions frequency at country level and during the period 2004–2017.

86 WFP. 2015. *Impact of climate related shocks and stresses on nutrition and food security in selected areas of rural Bangladesh*. Rome.

87 J. Hoddinott and B. Kinsey. 2001. Child growth in the time of drought. *Oxford Bulletin of Economics and Statistics*, 63(4): 409–436.

88 K. Grace, F. Davenport, C. Funk and A.M. Lerner. 2012. Child malnutrition and climate in sub-Saharan Africa: An analysis of recent trends in Kenya. *Applied Geography*, 35: 405–413.

89 Holleman, Rembold and Crespo (forthcoming). (see note 68).

90 PoU is a complex, aggregate measure of undernourishment at country level and is analysed from different perspectives in order to detect associations between climate variability/extremes and PoU. Based on the analysis presented, climate variability and extremes, including the 2015–2016 El Nino event, appear to be contributing factors for some countries, as seen both through the lens of climatology, as well as the PoU analysis of change points in correspondence of severe drought stress conditions. Figures 26 and 27 descriptively complement this analysis showing trends in PoU by different vulnerability categories related to climate-sensitive production and yield, severe drought food security sensitivity and dependence on agriculture. In addition to climate, however, there may be other factors at play, synthesized in the PoU, that affect undernourishment during this period.

91 FSIN, 2018 (see note 62). The reported food insecurity crisis-level population estimates are derived from a selection of countries and/or population groups that faced a high severity and magnitude of acute food insecurity in the period from January to December 2016. The key information source is the Integrated Food Security Phase Classification (IPC)/ Cadre Harmonisé (CH), which considers: countries with any segment of the population in IPC/CH Phase 4 Emergency or Phase 5 Catastrophe; countries with at least 1 million people

in IPC/CH Phase 3 Crisis; and countries for which an Inter-Agency Standing Committee (IASC) Humanitarian System-Wide Emergency Response was declared. These numbers represent emergency-level food insecurity that requires immediate humanitarian action. They differ from the prevalence of undernourishment estimates presented in previous sections, which are more globally comprehensive and measure chronic food deprivation.

92 FSIN, 2018 (see note 62).

93 J.M. Rodriguez-Llanes, S. Ranjan-Dash, O. Degomme, A. Mukhopadhyay and D. Guha-Sapir. 2011. Child malnutrition and recurrent flooding in rural eastern India: a community-based survey. *BMJ Open*, 1: e000109

94 R.K. Phalkey, C. Aranda-Jan, S. Marx, B. Höfle and R. Sauerborn. 2015. Systematic review of current efforts to quantify the impacts of climate change on undernutrition. *Proceedings of the National Academy of Sciences*, 112(33): E4522.

95 For a full analysis of conflict as a driver of increasing food security see: FAO, IFAD, UNICEF, WFP and WHO. 2017. *The State of Food Security and Nutrition in the World 2017. Building resilience for peace and food security*. Rome, FAO.

96 FAO, IFAD and WFP. 2015. *The State of Food Insecurity in the World 2015. Meeting the 2015 international hunger targets: taking stock of uneven progress*. Rome, FAO.

97 FAO, IFAD, UNICEF, WFP and WHO, 2017 (see note 95).

98 J.R. Porter, L. Xie, A.J. Challinor, K. Cochrane, S.M. Howden, M.M. Iqbal, D.B. Lobell, and M.I. Travasso. 2014. Food security and food production systems. In IPCC. 2014. *Climate Change 2014: Impacts, Adaptation, and Vulnerability. Part A: Global and Sectoral Aspects. Contribution of Working Group II to the Fifth Assessment Report of the Intergovernmental Panel on Climate Change*, pp. 485-533 [C.B. Field, V.R. Barros, D.J. Dokken, K.J. Mach, M.D. Mastrandrea, T.E. Bilir, M. Chatterjee, K.L. Ebi, Y.O. Estrada, R.C. Genova, B. Girma, E.S. Kissel, A.N. Levy, S. MacCracken, P.R. Mastrandrea and L.L. White, eds]. Cambridge, United Kingdom and New York, NY, USA, Cambridge University Press.

99 Porter *et al.*, 2014 (see note 98).

100 J. Hansen, S. Mason, L. Sun and A. Tall. 2011. Review of seasonal climate forecasting for agriculture in sub-Saharan Africa. *Experimental Agriculture, 47*(2): 205–240; T. Iizumi, J. Luo, A.J. Challinor, G. Sakurai, M. Yokozawa, H. Sakuma, M.E. Brown and T. Yamagata. 2014. Impacts of El Nino Southern Oscilation on the global yields of major crops. *Nature Communications, 5.*

101 M. Zampieri, A. Ceglar, F. Dentener and A. Toreti. 2017. Wheat yield loss attributable to heat waves, drought and water excess at the global, national and subnational scales. *Environmental Research Letters, 12*(6).

102 D.K. Ray, J.S. Gerber, G.K. MacDonald and P.C. West. 2015. Climate variation explains a third of global crop yield variability. *Nature Communications, 6.*

103 Porter *et al.*, 2014 (see note 98).

104 Porter *et al.*, 2014 (see note 98); M. Matiu, D.P. Ankerst and A. Menzel. 2017. Interactions between temperature and drought in global and regional crop yield variability during 1961–2014. *PloS ONE, 12*(5).

105 A.L. Hoffman, A.R. Kemanian and C.E. Forest. 2017. Analysis of climate signals in the crop yield record of sub-Saharan Africa. *Global Change Biology, 24*(1): 143–157.

106 T. Garg, M. Jagnani and V. Taraz. 2017. *Human Capital Costs of Climate Change: Evidence from Test Scores in India.* San Diego, USA, University of California.

107 M. Niles, J. Esquivel, R. Ahuja and N. Mango. 2017. *Climate: Change and Food Systems: Assessing Impacts and Opportunities.* Washington, DC, Meridian Institute.

108 Hansen *et al.*, 2011 (see note 100); Iizumi *et al.*, 2014 (see note 100).

109 T. Iizumi and N. Ramankutty. 2015. How do weather and climate influence cropping area and intensity? *Global Food Security, 4*(2015): 46–50.

110 G. Rabbani, A. Rahman and K. Mainuddin. 2013. Salinity-induced loss and damage to farming households in coastal Bangladesh. *International Journal of Global Warming, 5*(4): 400–415.

111 Iizumi and Ramankutty, 2015 (see note 109).

112 T. Sakamoto, N.V. Nguyen, H. Ohno, N.Ishitsuka and M. Yokozawa. 2006. Aptio-temporal distribution of rice phenology and cropping systems in Mekong Delta with special reference to the seasonal water flow of the Mekong and Bassac rivers. *Remote Sensing of Environment,* 100: 1–16.

113 Other case studies include:
Brazil – A.S. Cohn, L.K. VanWey, S.A. Spera and J.F. Mustard. 2016. Cropping frequency and area response to climate variability can exceed yield response. *Nature Climate Change,* 6: 601–604; sub-Saharan Africa – K. Waha, C. Müller and S. Rolinski. 2013. Separate and combined effects of temperature and precipitation change on maize yields in sub-Saharan Africa from mid to late 21st century. *Global and Planetary Change* 106: 1–12; India – S. Mondal, R.P. Singh, E.R. Mason, J. Huerta-Espino, E. Autrique and A.K. Joshi. 2016. Grain yield, adaptation and progress in breeding for early-maturing and heat-tolerant wheat lines in South Asia. *Field Crops Research Volume* 192: 78–85; Philippines – A.W. Robertson, A.V.M. Ines, J. Qian, D.G. DeWitt, A. Lucero and N. Koide. 2013. Prediction of rice production in the Philippines using seasonal climate forecasts. *Journal of Applied Meteorology and Climatology,* 52(3): 552–569.

114 K. Lewis. 2017. Understanding climate as a driver of food insecurity in Ethiopia. *Climatic Change,* 144(2): 317–328.

115 K. Lewis, 2017 (see note 114).

116 Codjoe and Owusu, 2011 (see note 70).

117 T. Wei, S. Glomsrød and T. Zhang. 2017. Extreme weather, food security and the capacity to adapt – the case of crops in China. *Food Security,* Volume 9(3): 523–535.

118 P. Lehodey, J. Alheit, M. Barange, T. Baumgartner, G. Beaugrand, K. Drinkwater, J.-M. Fromentin *et al.* 2006. Climate Variability, Fish and Fisheries. *Journal of Climate,* 19: 5009–5030.

119 FAO. 2018. *The impact of disasters and crises on agriculture and food security 2017.* Rome.

120 FAO, 2018 (see note 119).

121 FAO, 2015 (see note 60).

122 FAO. 2016. *Dry Corridor Central America Situation Report – June 2016*. Rome.

123 Food and Nutrition Security Working Group (FSNWG). 2016. *Southern Africa Food and Nutrition Security Update*.

124 Southern African Development Community (SADC). 2016. *Regional situation Update on the El Nino-Induced Drought*. Issue 2 and 3 [online]. Gaborone. www.sadc.int/news-events/newsletters/sadc-regional-situation-update-el-nino-induced-drought

125 Holleman, Rembold and Crespo (forthcoming). (see note 68).

126 FAO, 2015 (see note 60).

127 FAO, 2015 (see note 60).

128 FAO, 2018 (see note 119).

129 FAO, 2015 (see note 60).

130 FAO, 2015 (see note 60).

131 Agriculture value added is the net output of the agriculture sector and subsectors after adding all outputs and subtracting intermediate inputs. Agriculture value-added growth is the annual percentage change of agriculture value added. See FAO, 2015 (see note 60).

132 Global Panel on Agriculture and Food Systems for Nutrition. 2016. *Food systems and diets: Facing the challenges of the 21st century*. London.

133 T. Wheeler and J. von Braun. 2013. Climate change impacts on global food security. *Science*, 341(6145): 508–513.

134 C. Béné, J. Waid, M. Jackson-deGraffenried, A. Begum, M. Chowdhury, V. Skarin, A. Rahman, N. Islam, N. Mamnun, K. Mainuddin and S.M.A. Amin. 2015. *Impact of climate-related shocks and stresses on nutrition and food security in selected areas of rural Bangladesh*. Dhaka, WFP.

135 Holleman, Rembold and Crespo (forthcoming). (see note 68).

136 Holleman, Rembold and Crespo (forthcoming). (see note 68).

137 M. Peri. 2017. Climate variability and the volatility of global maize and soybean prices. *Food Security*, 9(4): 673–683.

138 FAO. 2016. *The State of Food and Agriculture 2016: climate change, agriculture and food security*. Rome.

139 FAO, International Fund for Agricultural Development (IFAD), International Monetary Fund (IMF), Organization for Economic Cooperation and Development (OECD), United Nations Conference on Trade and Development (UNCTAD), WFP, World Bank, the World Trade Organization (WTO), International Food Policy Research Institute (IFPRI) and the High Level Task Force on Global Food and Nutrition Security (HLTF). 2011. *Price Volatility in Food and Agriculture Markets: Policy Responses*.

140 FAO, 2016 (see note 138).

141 G. Rapsomanikis. 2015. *The economic lives of smallholder farmers. An analysis based on household data on nine countries*. Rome, FAO.

142 Met Office and WFP. 2012. *Climate impacts on food security and nutrition. A review of existing knowledge*. Devon, UK and Rome; M. Brown and C. Funk. 2008. Food security under climate change. *NASA Publications*, 319(5863): 580–581.

143 S. Asfaw and G. Maggio. 2018. Gender, weather shocks and welfare: evidence from Malawi. *Journal of Development Studies*, 54(2): 271–291; M. Asfaw, M. Wondaferash, M. Taha and L. Dube. 2015. Prevalence of undernutrition and associated factors among children aged between six to fifty nine months in Bule Hora district, South Ethiopia. *BMC Public Health*, 15(41).

144 FAO, 2016 (see note 55).

145 FAO, 2016 (see note 138).

146 H. Alderman. 2010. Safety nets can help address the risks to nutrition from increasing climate variability. *The Journal of Nutrition*, 140(1): 148S–152S; M.T. Ruel, H. Alderman and the Maternal and Child Nutrition Study Group. 2013. Nutrition-sensitive interventions and programmes: how can they help to accelerate progress in improving maternal and child nutrition? *The Lancet*, 382(9891): 536–551.

147 A.L. Thorne-Lyman, N. Valpiani, K. Sun, R.D. Semba, C.L. Klotz, K. Kraemer, N. Akhter, S. de Pee, R. Moench-Pfanner, M. Sari and M.W. Bloem. 2009. Household dietary diversity and food expenditures are closely linked in rural Bangladesh, increasing the risk of malnutrition due to the financial crisis. *The Journal of Nutrition*, 140(1): 182S–188S; H. Torlesse, L. Kiess and M.W. Bloem. Association of household rice expenditure with child nutritional status indicates a role for macroeconomic food policy in combating malnutrition. *The Journal of Nutrition*, 133(5): 1320–1325.

148 M. Sari, S. de Pee, M.W. Bloem, K. Sun, A.L. Thorne-Lyman, R. Moench-Pfanner, N. Akhter, K. Kraemer and R.D Semba. 2009. Higher household expenditure on animal-source and nongrain foods lowers the risk of stunting among children 0–59 months old in Indonesia: implications of rising food prices. *Journal of Nutrition*, 140(1): 195S–200S.

149 N.J. Saronga, I.H. Mosha, A.T. Kessy, M.J. Ezekiel, A. Zizinga, O. Kweka, P. Onyango and S. Kovats. 2016. "I eat two meals per day" impact of climate variability on eating habits among households in Rufiji district, Tanzania: a qualitative study. *Agriculture and Food Security*, 5(14).

150 B. Vaitla, S. Devereux and S.H. Swan. 2009. Seasonal hunger: a neglected problem with proven solutions. *PloS Medicine*, 6(6): e1000101; G. Egata, Y. Berhane and A. Worku. 2013. Seasonal variation in the prevalence of acute undernutrition among children under five years of age in east rural Ethiopia: a longitudinal study. *BMC Public Health*, 13(864); IFPRI. 2015. *Global Nutrition Report 2015: Actions and Accountability to Advance Nutrition and Sustainable Development*. Washington, DC.

151 B.R. Guzman Herrador, B. Freiesleben de Blasio, E. MacDonald, G. Nichols, B. Sudre, L. Vold, J.C. Semenza and K. Nygård. 2015. Analytical studies assessing the association between extreme precipitation or temperature and drinking water-related waterborne infections: a review. *Environmental Health*, 14(29); Z. Herrador, J. Perez-Formigo, L. Sordo, E. Gadisa, J. Moreno, A. Benito, A. Aseffa and E. Custodio. 2015. Low dietary diversity and intake of animal source foods among school aged children in Libo Kemkem and Fogera Districts, Ethiopia. *PloS One* 205, 10(7): e0133435; B.G. Luckett, F.A. DeClerck, J. Fanzo, A.R. Mundorf and D. Rose. 2015. Application of the nutrition functional diversity indicator to assess food system contributions to dietary diversity and sustainable diets of Malawian households. *Public Health Nutrition*, 18(13):

2479–2487; J.E. Ntwenya, J. Kinabo, J. Msuya, P. Mamiro and Z.S. Majili. 2015. Dietary patterns and household food insecurity in rural populations of Kilosa District, Tanzania. *PloS One*, 10(5): e0126038; F.K. M'Kaibi, N.P. Steyn, S. Ochola and L. Du Plessis. 2015. Effects of agricultural biodiversity and seasonal rain on dietary adequacy and household food security in rural areas of Kenya. *BMC Public Health*, 15(422); K.T. Roba, T.P. O'Connor, T. Belachew and N.M. O 'Brien. 2016. Variations between post- and pre-harvest seasons in stunting, wasting, and Infant and Young Child Feeding (IYCF) practices among children 6-23 months of age in lowland and midland agro-ecological zones of rural Ethiopia. *Pan African Medical Journal*, 24(163); M. Mayanja, M.J. Rubaire-Akiiki, S. Young and T. Greiner. 2015. Diet diversity in pastoral and agro-pastoral households in Ugandan rangeland ecosystems. *Ecology of Food and Nutrition*, 54(5): 529–545; M. Stelmach-Mardas, C. Kleiser, I. Uzhova, J.L. Peñalvo, G. La Torre, W. Palys, D. Lojko, K Nimptsch, A. Suwalska, J. Linseisen, R. Saulle, V. Colamesta and H. Boeing. 2016. Seasonality of food groups and total energy intake: a systematic review and meta-analysis. *European Journal of Clinical Nutrition*, 70(6): 700–708.

152 Seasonality and seasonal income insecurity is a feature of poverty in many parts of the world. In agricultural regions of developing countries, it is known as the lean season – that dangerous period between planting and harvesting when job opportunities are scarce and incomes plummet. The lean season is associated with low income and hunger. A. Gelli, N. Aberman, A. Margolies, M. Santacroce, B. Baulch and E. Chirwa. 2017. Lean-season food transfers affect children's diets and household food security: evidence from a quasi-experiment in Malawi. *The Journal of Nutrition*, 147(5): 869–878.

153 L.T. Huong, L.T.T. Xuan, L.H. Phuong, D.T.T. Huyen and J. Rocklöv. 2014. Diet and nutritional status among children 24–59 months by seasons in a mountainous area of Northern Vietnam in 2012. *Global Health Action*, 7(23121).

154 A. Seiden, N.L. Hawley, D. Schulz, S. Raifman and S.T. McGarvey. 2012. Long-Term Trends in Food Availability, Food Prices, and Obesity in Samoa. *American Journal of Human Biology*, 24(3): 286–95; J. Campbell. 2015. Development, global change and traditional food security in Pacific Island countries. *Regional Environmental Change*, 15(7): 1313–24.

155 T. Stathers, R. Lamboll, B.M. Mvumi. 2013. Postharvest agriculture in changing climates: its importance to African smallholder farmers. *Food Security,* 5(3): 361–392.

156 World Health Organization (WHO). 2017. Food Safety: Fact sheet No 399 [online]. Geneva, Switzerland. www.who.int/mediacentre/factsheets/fs399

157 WHO. 2015. *WHO estimates of the global burden of foodborne diseases: foodborne disease burden epidemiology reference group 2007–2015.* Geneva, Switzerland.

158 Similarly, cases of salmonellosis increased by 5–10 percent for each 1 °C increase in weekly temperature when ambient temperatures are above 5 °C in Europe. See WHO. 2017. *Protecting health in Europe from climate change: 2017 update* [online]. Copenhagen. www.euro.who.int/__data/assets/pdf_file/0004/355792/ProtectingHealthEurope FromClimateChange.pdf?ua=1

159 S. Moniruzzaman. 2015. Crop choice as climate change adaptation: Evidence from Bangladesh. *Ecological Economics,* 118: 90–98.

160 K.R. Smith, A. Woodward, D. Campbell-Lendrum, D.D. Chadee, Y. Honda, Q. Liu, J.M. Olwoch, B. Revich and R. Sauerborn. 2014. Human health: impacts, adaptation, and co-benefits. In IPCC. 2014. *Climate Change 2014: Impacts, adaptation, and vulnerability. Contribution of Working Group II to the Fifth Assessment Report of the Intergovernmental Panel on Climate Change,* pp. 709–754 [C.B. Field, V.R. Barros, D.J. Dokken, K.J. Mach, M.D. Mastrandrea, T.E. Bilir, M. Chatterjee, K.L. Ebi, Y.O. Estrada, R.C. Genova, B. Girma, E.S. Kissel, A.N. Levy, S. MacCracken, P.R. Mastrandrea and L.L. White, eds]. Cambridge, UK, and New York, USA, Cambridge University Press.

161 Smith *et al.,* 2014 (see note 160).

162 N. Watts, M. Ammann, S. Ayeb-Karlsson, K. Belesova, T. Bouley, M. Boykoff, P.Byass, *et al.* 2016. The *Lancet* Countdown on health and climate change: from 25 years of inaction to a global transformation for public health. *The Lancet,* 391(10120): 581–630.

163 G.P. Kenny, J. Yardley, C. Brown, R.J. Sigal and O. Jay. 2010. Heat stress in older individuals and patients with common chronic diseases. *Canadian Medical Association Journal,* 182(10): 1053–1060.

164 Watts *et al.,* 2016 (see note 162).

165 N. Watts, M. Amann, S. Ayeb-Karlsson, K. Belesova, T. Bouley, M. Boykoff, P. Byass *et al.* 2018. The *Lancet* Countdown on health and climate change: from 25 years of inaction to a global transformation for public health. *The Lancet,* 391(10120): 581–630.
The labour capacity was defined as labour capacity=100–25-max (0, WBGT–25)2/3 where WBGT is the Wet Bulb Globe Temperature, which is a 17 function of the dew point temperature (see J.P. Dunne, R.J. Stouffer and J.G. John. 2013. Reductions in labour capacity from heat stress under climate warming. *Nature Climate Change,* 3: 563–566). See N. Watts, M. Amann, S. Ayeb-Karlsson, K. Belesova, T. Bouley, M. Boykoff, P. Byass *et al.* 2017. Supplement to the *Lancet* Countdown on health and climate change: from 25 years of inaction to a global transformation for public health. *The Lancet,* 391(10120).

166 Smith *et al.,* 2014 (see note 160).

167 T.G. Veenema, C.P. Thornton, R.P. Lavin, A.K. Bender, S. Seal and A. Corley. 2017. Climate change-related water disasters' impact on population health. *Journal of Nursing Scholarship,* 49(6): 625–634.

168 K.F. Cann, D.R. Thomas, R.L. Salmon, A.P. Wyn-Jones and D. Kay. 2013. Extreme water-related weather events and waterborne disease. *Epidemiology and Infection,* 141(4): 671–686.

169 J.P. Chretien, A. Anyamba, J. Small, S.Britch, J. L. Sanchez, A.C. Halbach, C. Tucker and K. J. Linthicum. 2015. Global Climate Anomalies and Potential Infectious Disease Risks, *PLoS Currents,* 7.

170 K. Brown. 2003. Diarrhea and malnutrition. *Journal of Nutrition,* V133(1): 328S–332S.

171 M. Azage, A. Kumie, A. Worku, A.C. Bagtzoglou and E. Anagnostou. 2017. Effect of climatic variability on childhood diarrhea and its high risk periods in northwestern parts of Ethiopia. *PLoS One,* 12(10): e0186933.

172 Development Initiatives. 2017. *Global Nutrition Report 2017: Nourishing the SDGs.* Bristol, UK. According to this report, in Cambodia: one in four children under five is underweight, one in ten suffers from wasting, and one in three is stunted.

173 WHO and World Meteorological Organization (WMO). 2012. *Atlas of Health and Climate.* Geneva, Switzerland, WHO Press.

174 Smith *et al.*, 2014 (see note 160).

175 R.S. Kovats, M.J. Bouma, S. Hajat, E. Worrall and A. Haines. 2003. El Niño and health. *Lancet*, 362(9394): 1481–1489; S.M. Moore, A.S. Azman, B.F. Zaitchik, E. D. Mintz, J. Brunkard, A. Legros, A. Hill, H. McKay, F.J. Luquero, D. Olson and J. Lesslera. 2017. El Niño and the shifting geography of cholera in Africa. *Proceedings of the National Academy of Sciences of the United States of America*, 114(17): 4436–4441.

176 See T. Gone, F. Lemango, E. Eliso, S. Yohannes and T. Yohannes. 2017. The association between malaria and malnutrition among under-five children in Shashogo District, Southern Ethiopia: a case-control study. *Infectious Diseases of Poverty*, 6(9); B. Shikur, W. Deressa and B. Lindtjørn. 2016. Association between malaria and malnutrition among children aged under-five years in Adami Tulu District, south-central Ethiopia: a case–control study. *BMC Public Health*, 16(174); M.A. Araújo Alexandre, S. Gomes Benzecry, A. Machado Siqueira, S. Vitor-Silva, G. Cardoso Melo, W.M. Monteiro, H. Pons Leite, M.V. Guimarães Lacerda and M. Costa Alecrim. 2015. The Association between Nutritional Status and Malaria in Children from a Rural Community in the Amazonian Region: A Longitudinal Study. *PloS Neglected Tropical Diseases*, 9(4): e0003743; C. E. Oldenburg, P. J. Guerin, F. Berthé, R. F. Grais and S. Isanaka. 2018. Malaria and Nutritional Status Among Children With Severe Acute Malnutrition in Niger: A Prospective Cohort Study. *Clinical Infectious Diseases*, ciy207.

177 WHO. 2017. Malaria in pregnant women. In *WHO* [online]. Geneva, Switzerland. www.who.int/malaria/areas/high_risk_groups/pregnancy/en

178 WHO and WMO, 2012 (see note 173).

179 Smith *et al.*, 2014 (see note 160).

180 N. Watts, W.N. Adger, P. Agnolucci, J. Blackstock, P. Byass, W. Cai, S. Chaytor *et al.* 2015. Health and climate change: policy responses to protect public health. *The Lancet*, 386(10006): 1861–1914.

181 WHO, 2009. Protecting health from climate change: connecting science, policy and people. Geneva, Switzerland.

182 H. Frumkin, J. Hess, G. Luber, J. Malilay and M. McGeehin. 2008. Climate Change: The Public Health Response. *American Journal of Public Health*; 98(3): 435–445; Smith *et al.*, 2014 (see note 160).

183 B. Campbell, S. Mitchell and M. Blackett. 2009: *Responding to Climate Change in Vietnam. Opportunities for Improving Gender Equality.* A Policy Discussion Paper. Hanoi, Oxfam and UN.

184 C. S. Homer, E. Hanna and A.J. McMichael. 2009. Climate change threatens the achievement of the millennium development goal for maternal health. *Midwifery*, 25(6): 606–612.

185 Oxfam International. 2005. *Oxfam Briefing Note: The tsunami's impact on women.* Oxford, UK.

186 Y. Lambrou and S. Nelson. 2013. Gender issues in climate change adaptation: farmers' food security in Andhra Pradesh. In M. Alston and K. Whittenbury, eds. *Research, Action and Policy: Addressing the Gendered Impacts of Climate Change*, pp. 189–206. Dordrecht, Netherlands, Springer Science.

187 S. Neelormi, N. Adri and A. Uddin Ahmed. 2008. *Gender Perspectives of Increased Socio-Economic Risks of Waterlogging in Bangladesh due to Climate Change.* St. Petersburg, USA, International Ocean Institute; Campbell *et al.*, 2009 (see note 183).

188 U.T. Okpara, L.C. Stringer and A.J. Dougill. 2016. Lake drying and livelihood dynamics in Lake Chad: Unravelling the mechanisms, contexts and responses. *Ambio*, 45(7): 781–795.

189 A.D. Jones, Y. Cruz Agudo, L. Galway, J. Bentley and P. Pinstrup-Andersen. 2012. Heavy agricultural workloads and low crop diversity are strong barriers to improving child feeding practices in the Bolivian Andes. *Social Science & Medicine*, 75(9): 1673–1684; L. Olsson, M. Opondo, P. Tschakert, A. Agrawal, S.H. Eriksen, S. Ma, L.N. Perch and S.A. Zakieldeen. 2014. Livelihoods and poverty. In IPCC. 2014. *Climate Change 2014: Impacts, Adaptation, and Vulnerability. Part A: Global and Sectoral Aspects. Contribution of Working Group II to the Fifth Assessment Report of the Intergovernmental Panel on Climate Change*, pp. 793–832 [C.B. Field, V.R. Barros, D.J. Dokken, K.J. Mach, M.D. Mastrandrea, T.E. Bilir, M. Chatterjee, K.L. Ebi, Y.O. Estrada, R.C. Genova, B. Girma, E.S. Kissel, A.N. Levy, S. MacCracken, P.R. Mastrandrea and L.L. White,

eds]. Cambridge, UK, and New York, USA, Cambridge University Press.

190 United Nations System Standing Committee on Nutrition (UNSCN). 2010. *Climate change and nutrition security.* Geneva, Switzerland; Jones *et al.*, 2012 (see note 189).

191 A. Datar, J. Liu, S. Linnemayr and C. Stecher. 2013. The impact of natural disasters on child health and investments in rural India. *Social Science & Medicine*, 76(1): 83–91.

192 J. Fanzo, R. McLaren, C. Davis and J. Choufani. 2017. *Climate change and variability. What are the risks for nutrition, diets, and food systems?* Washington, DC.

193 International Union for Conservation of Nature (IUCN), International Institute for Sustainable Development (IISD), Stockholm Environment Institute, Swiss Agency for Development and Cooperation (SDC) and Swiss Organisation for Development and Cooperation (Intercooperation). 2003. *Livelihoods and Climate Change: Combining disaster risk reduction, natural resource management and climate change adaptation in a new approach to the reduction of vulnerability and poverty: A Conceptual Framework Paper Prepared by the Task Force on Climate Change, Vulnerable Communities and Adaptation.* Winnipeg, Canada, IISD; M.-C. Badjeck, E.H. Allison. A.S.Halls and N.K. Dulvyef. 2010. Impacts of climate variability and change on fishery-based livelihoods. *Marine Policy*, 34(3): 375–383.

194 FAO, 2015 (see note 60).

195 FAO, 2015 (see note 60); FAO. 2013. *Resilient livelihoods. Disaster Risk Reduction for Food and Nutrition Security.* Rome; FAO, 2018 (see note 119); UNSCN. 2016. *Impact Assessment of Policies to support Healthy Food Environments and Healthy Diets - Implementing the Framework for Action of the Second International Conference on Nutrition.*

196 FAO, 2018 (see note 119).

197 FAO, 2018 (see note 119).

198 Oxfam International. 2011. *Pakistan Floods Progress Report - July 2010 / July 2011.* Oxford, UK; P.K. Krishnamurthy, K. Lewis, C. Kent and P. Aggarwal. 2015. *Climate impacts on food security and livelihoods in Asia: A review of existing knowledge.* Bangkok, WFP Regional Bureau for Asia; Devon, UK, Met Office UK; and New Delhi, CGIAR–CCAFS International Water Management Institute.

199 A. Elbehri, A. Challinor, L. Verchot, A. Angelsen, T. Hess, A. Ouled Belgacem, H. Clark, *et al.* 2017. *FAO-IPCC Expert Meeting on Climate Change, Land Use and Food Security: Final Meeting Report; January 23–25, 2017.* Rome, FAO and IPCC.

200 Global Panel on Agriculture and Food Systems for Nutrition. 2016. *Food systems and diets: Facing the challenges of the 21st century.* London.

201 CRED, 2015 (see note 59).

202 J.M. Rodriguez-Llanes, S. Ranjan-Dash, O. Degomme, A. Mukhopadhyay and D. Guha-Sapir. 2011. Child malnutrition and recurrent flooding in rural eastern India: a community-based survey. *BMJ Open, 1: e000109.*

203 FSNWG, 2016 (see note 123).

204 J. Hesselberg and J.A. Yaro. 2006. An assessment of the extent and causes of food insecurity in northern Ghana using a livelihood vulnerability framework. *GeoJournal*, 67(1): 41–55; J.A. Yaro. 2006. Is deagrarianisation real? A study of livelihood activities in rural northern Ghana. *The Journal of Modern African Studies*, 44(1): 125–156; Codjoe and Owusu, 2011 (see note 70); L. Connolly-Boutin and B. Smit. 2016. Climate change, food security, and livelihoods in sub-Saharan Africa. *Regional Environmental Change*, 16(2): 385–399.

205 Badjeck *et al.*, 2010 (see note 193).

206 F.R. Sansoucy. 1995. Livestock – a driving force for food security and sustainable development. In J. Diouf. *World Animal Review.* Rome, FAO.

207 T. Schillhorn van Veen. 2001. *Livestock-in-kind credit: helping the rural poor to invest and save.* Washington, DC, World Bank.

208 FAO. 2017. *Somalia 2017: Saving livestock, saving livelihoods and saving lives.* Rome.

209 M.R. Carter, P.D. Little, T. Mogues and W. Negatu. 2007. Poverty traps and natural disasters in Ethiopia and Honduras. *World Development*, 35(5): 835–856; J. Hoddinott. 2006. Shocks and their consequences across and within households in rural Zimbabwe. *The Journal of Development Studies*, 42(2): 301–321.

210 Badjeck *et al.*, 2010 (see note 193).

211 Badjeck *et al.*, 2010 (see note 193).

212 S. Prakash. 2002. Social capital and the rural poor: what can civil actors and policies do? In *Social Capital and Poverty Reduction: Which role for civil society, organizations and the State?*, pp. 49–62. Paris, UNESCO.

213 FAO, IFAD, UNICEF, WFP and WHO, 2017 (see note 95).

214 M. Burke, S. Hsiang and E. Miguel. 2015. Climate and Conflict. *Annual Review of Economics*, 7: 577–617.

215 Holleman, Rembold and Crespo (forthcoming). (see note 68).

216 See, for example, FAO, IFAD, UNICEF, WFP and WHO, 2017 (see note 95).

217 C. del Ninno, P. Dorosh and L. Smith. 2003. *Public Policy, Food Markets, and Household Coping Strategies in Bangladesh: Lessons from the 1998 Floods.* Food Consumption and Nutrition Division Paper No.156. Washington, DC, IFPRI.

218 B.N. Nwokeoma and A.K. Chinedu. 2017. Climate Variability and Consequences for Crime, Insurgency in North East Nigeria. *Mediterranean Journal of Social Sciences*, 8(3): 171–182.

219 WFP, IOM and LSE. 2015. *Hunger without Borders, the hidden links between food insecurity, violence and migration in the northern triangle of Central America.*

220 Hansen *et al.*, 2011 (see note 100).

221 C. Elbers, J.W. Gunning and B. Kinsey. 2007. Growth and risk: methodology and micro evidence. *World Bank Economic Review*, 21(1): 1–20.

222 S. Hallegatte, L. Bangalore, L. Bonzanigo, M. Fay, T. Kane, U. Narloch, J. Rozenberg *et al.* 2016. *Shock Waves: Managing the Impacts of Climate Change on Poverty.* Climate Change and Development. Washington, DC, World Bank.

223 M. Rosenzweig and K.I. Wolpin. 1993. Credit Market Constraints, Consumption Smoothing, and the Accumulation of Durable Production Assets in Low-Income Countries: Investment in Bullocks in India. *Journal of Political Economy*,

101(2): 223–244; M. Fafchamps, C. Udry and K. Czukas. 1998. Drought and saving in West Africa: are livestock a buffer stock? *Journal of Development Economics*, 55(2): 273–305; H. Kazianga and C. Udry. 2006. Consumption smoothing? Livestock, insurance and drought in rural Burkina Faso. *Journal of Development Economics*, 79(2): 413–446; A.R. Quisumbing. 2008. *Intergenerational transfers and the intergenerational transmission of poverty in Bangladesh: Preliminary results from a longitudinal study of rural households.* Chronic Poverty Research Centre Working Paper No. 117. Manchester, UK, University of Manchester.

224 M. Eswaran and A. Kotwal. 1990. Implications of Credit Constraints for Risk Behaviour in Less Developed Economies. *Oxford Economic Papers*, 42(2): 473–482; M. Rosenzweig and H. Binswanger. 1993. Wealth, Weather Risk and the Composition and Profitability of Agricultural Investments. *Economic Journal*, 103(416): 56–78; F.J. Zimmerman and M. Carter. 2003. Asset smoothing, consumption smoothing and the reproduction of inequality under risk and subsistence constraints. *Journal of Development Economics*, 71(2): 233–260.

225 C.B. Barrett, C.M. Moser, O.V. McHugh and J. Barison. 2004. Better technology, better plots, or better farmers? Identifying changes in productivity and risk among Malagasy rice farmers. *American Journal of Agricultural Economics*, 86(4): 869–888; S. Dercon. 1996. Risk, crop choice, and savings: evidence from Tanzania. *Economic Development and Cultural Change*, 44(3): 485–513; M. Fafchamps. 2003. *Inequality and Risk.* Economics Series Working Papers 141. Oxford, UK, University of Oxford; Y. Kebede. 1992. Risk behaviour and new agricultural technologies: the case of producers in the Central Highlands of Ethiopia. *Quarterly, Journal of International Agriculture*, 31: 269–284; M. Marra, D.J. Pannell, A.A. Ghadim. 2003. The economics of risk, uncertainty and learning in the adoption of new agricultural technologies: where are we on the learning curve? *Agricultural Systems*, 75(2): 215–234. E. Rose. 2001. *Ex ante* and *ex post* labor supply response to risk in a low-income area. *Journal of Development Economics*, 64(2): 371–388; M.R. Rosenzweig and O. Stark. 1989. Consumption Smoothing, Migration, and Marriage: Evidence from Rural India. *Journal of Political Economy*, 97(4): 905–926.

226 E. Bryan, T.T. Deressa, G.A. Gbetibouo and C. Ringler. 2009. Adaptation to climate change in Ethiopia and South Africa: options and constraints. *Environmental Science & Policy*, 12(4): 413–426.

227 T.E. Downing, L. Ringius, M. Hulme and D. Waughray. 1997. Adapting to climate change in Africa. *Mitigation and Adaptation Strategies for Global Change*, 2: 19. L. Westerhoff and B. Smit. 2009. The rains are disappointing us: dynamic vulnerability and adaptation to multiple stressors in the Afram Plains, Ghana. *Mitigation and Adaptation Strategies for Global Change* 14: 317–337.

228 M. Casale, S. Drimie, T. Quinlan and G. Ziervogel. 2010. Understanding vulnerability in Southern Africa: comparative findings using a multiple-stressor approach in South Africa and Malawi. *Regional Environmental Change*, 10(2): 157–168; W. Laube, B. Schraven and M. Awo. 2012. Smallholder adaptation to climate change: Dynamics and limits in Northern Ghana, *Climate Change*, 111(3–4): 753–774; Tambo and Abdoulaye, 2013 (see note 72).

229 B. Smit and J. Wandel. 2006. Adaptation, adaptive capacity and vulnerability. *Global Environmental Change*, 16(3): 282–292.

230 S.T. Kandji, L. Verchot and J. Mackensen. 2006. *Climate change and variability in Southern Africa: Impacts and Adaptation in the agricultural sector.* Nairobi, United Nation Environmental Programme (UNEP) and World Agroforestry Centre (ICRAF).

231 S. Traerup and O. Mertz. 2011. Rainfall variability and household coping strategies in northern Tanzania. *Regional Environmental Change*, 11(3): 471–481; Tambo and Abdoulaye, 2013 (see note 72).

232 Yaro, 2006 (see note 204).

233 J.A. Tambo. 2016. Adaptation and resilience to climate change and variability in north-east Ghana. *International Journal of Disaster Risk Reduction*, 17: 85–94.

234 D.S.G. Thomas, C. Twyman, H. Osbahr and B. Hewitson. 2007. Adaptation to climate change and variability: farmer responses to intra-seasonal precipitation trends in South Africa. *Climatic Change*, 83(3): 301–322.

235 WFP. 2015. *More intense typhoons: What does a changing climate mean for food security in the Philippines?* Rome.

236 WFP. 2016. *Is the fun drying up? Implications of intensifying El Niño conditions for drought risk and food security.* Rome.

237 N.C.T. Castillo. 1990. Coping Mechanisms of Filipino Households in Different Agro-Ecological Settings. *Transactions of the National Academy of Science and Technology*, 12: 257–273.

238 Thomas *et al.*, 2007 (see note 234).

239 Tambo, 2016 (see note 233).

240 Tambo, 2016 (see note 233).

241 A. Arslan, R. Cavatassi, F. Alfani, N. Mccarthy, L. Lipper and M. Kokwe. 2017. Diversification under climate variability as part of CSA strategy in rural Zambia. *The Journal of Development Studies*, 54(3): 457–480; S. Asfaw, G. Pallante and A. Palma. 2018. Diversification Strategies and Adaptation Deficit: Evidence from Rural Communities in Niger. *World Development*, 101: 219–234.

242 Tambo, 2016 (see note 233).

243 Thomas *et al.*, 2007 (see note 234).

244 Z. Kubik and M. Maurel. 2016. Weather Shocks, Agricultural Production and Migration: Evidence from Tanzania, *The Journal of Development Studies*, 52(5): 665–680; Rosenzweig and Stark, 1989 (see note 225).

245 A. Agrawal and N. Perrin. 2009. Climate adaptation, local institutions, and rural livelihoods. *In* W.N. Adger, I. Lorenzoni and K.L. O'Brien, eds. *Adapting to climate change: Thresholds, values, governance*, pp 350–367. Cambridge, UK, Cambridge University Press; FAO, IFAD, UNICEF, WFP and WHO, 2017 (see note 95). See: K. Ober. 2014. *Migration as Adaptation: exploring mobility as a coping strategy for climate change*, UK Climate Change and Migration Coalition, Oxford, UK.

246 Agrawal and Perrin, 2009 (see note 245).

247 Rosenzweig and Stark, 1989 (see note 225).

248 Rosenzweig and Stark, 1989 (see note 225).

249 Kubik and Maurel, 2016 (see note 244).

250 Norwegian Refugee Council (NRC) and Internal Displacement Monitoring Centre (IDMC). 2015. *Global Estimates 2015: People displaced by disasters.* Châtelaine, Switzerland.

251 J. Barnett and M. Webber. 2010. *Accommodating Migration to Promote Adaptation to Climate Change.* Policy Research Working Paper 5270. New York, USA, World Bank;
E. Piguet, A. Pécoud and P. De Guchteneire, eds. 2011. Migration and Climate Change. Paris, UNESCO and Cambridge, UK, Cambridge University Press.

252 FAO (forthcoming). *The State of Food and Agriculture 2018. Migration, Agriculture and Rural Development.* Rome.

253 D. Maxwell and M. Fitzpatrick. 2012. The 2011 Somalia Famine: Context, Causes, and Complications. *Global Food Security,* 1(1): 5–12.

254 J. Hardoy and G. Pandiella. 2009. Urban poverty and vulnerability to climate change in Latin America. *Environment & Urbanization Copyright,* 21(1): 203–224.

255 H.C. Eakin and M.B. Wehbe. 2009. Linking local vulnerability to system sustainability in a resilience framework: two cases from Latin America. *Climatic Change,* 93(3–4): 355–377; W.E. Easterling. 2007. Climate change and the adequacy of food and timber in the 21st century. *Proceedings of the National Academy of Sciences of the United States of America,* 104(50): 19679; S. Eriksen and J.A. Silva. 2009. The vulnerability context of a savanna area in Mozambique: household drought coping strategies and responses to economic change. *Environmental Science & Policy,* 12(1): 33–52; IPCC. 2007. *Climate Change 2007: Synthesis Report. Contribution of Working Groups I, II and III to the Fourth Assessment Report of the Intergovernmental Panel on Climate Change.* [Core Writing Team, R.K. Pachauri and A. Reisinger, eds]. Geneva, Switzerland; IPCC. 2014. *Climate Change 2014: Impacts, Adaptation, and Vulnerability. Part A: Global and Sectoral Aspects. Contribution of Working Group II to the Fifth Assessment Report of the Intergovernmental Panel on Climate Change* [C.B. Field, V.R. Barros, D.J. Dokken, K.J. Mach, M.D. Mastrandrea, T.E. Bilir, M. Chatterjee, K.L. Ebi, Y.O. Estrada, R.C. Genova, B. Girma, E.S. Kissel, A.N. Levy, S. MacCracken, P.R. Mastrandrea and L.L. White, eds]. Cambridge, UK, and New York, USA, Cambridge University Press; J.F. Morton. 2007. The impact of climate change on smallholder and subsistence agriculture. *Proceedings of the National Academy of Sciences of the United States of America,* 104(50): 19680–19685; K. O'Brien, L. Sygna, R. Leichenko, N. Adger, J. Barnett, T. Mitchell, L. Schipper, T. Tanner, C. Vogel and C. Mortreux. 2008. *Disaster Risk Reduction, Climate Change Adaptation and Human Security:*

A Commissioned Report for the Norwegian Ministry of Foreign Affairs. Report GECHS Report 2008:3. Oslo, University of Oslo; P. Reid and C. Vogel. 2006. Living and responding to multiple stressors in South Africa—Glimpses from KwaZulu-Natal. *Global Environmental Change,* 16(2): 195–206; L. Schipper and M. Pelling. 2006. Disaster risk, climate change and international development: scope for, and challenges to, integration. *Disasters,* 30(1): 19–38. P. Tschakert. 2007. Views from the vulnerable: understanding climatic and other stressors in the Sahel. *Global Environmental Change,* 17(3–4): 381–396; G. Ziervogel, P. Johnston, M. Matthew and P. Mukheibir. 2010. Using climate information for supporting climate change adaptation in water resource management in South Africa. *Climatic Change,* 103(3–4): 537–554.

256 J. Hoddinott. 2006. Shocks and their Consequences across and within Households in Rural Zimbabwe. *The Journal of Development Studies,* 42(2): 301–321.

257 J. Barnett and S. O'Neill. 2010. Maladaptation. *Global Environmental Change,* 20(2): 211–213; T. Tanner and T. Mitchell. 2008. Introduction: Building the case for pro-poor adaptation. *IDS Bulletin,* 39(4): 1–5. Brighton, UK, Institute of Development Studies; G. Ziervogel, S. Bharwani and T.E. 2006. Adapting to climate variability: Pumpkins, people and policy. *Natural Resource Forum,* 30(4): 294–305.

258 J. Ribot. 2010. Vulnerability does not fall from the sky: toward multiscale, pro-poor climate policy. In R. Mearns and A. Norton, eds. *Social Dimensions of Climate Change: Equity and Vulnerability in a Warming World.* Washington, DC, The International Bank for Reconstruction and Development and World Bank.

259 R. Mearns and A. Norton. 2010. *Social Dimensions of Climate Change: Equity and Vulnerability in a Warming World.* New Frontiers of Social Policy. Washington, DC, World Bank.

260 Olsson *et al.,* 2014 (see note 189).

261 S. Hallegatte, A. Vogt-Schilb, M. Bangalore and J. Rozenberg. 2017. *Unbreakable: Building the Resilience of the Poor in the Face of Natural Disasters.* Climate Change and Development Series. Washington, DC, World Bank; M. Madajewicz and A.H. Tsegay. 2017. Managing Risks in Smallholder Agriculture: The Impacts of R4 on Livelihoods in Tigray, Ethiopia. Boston, Oxfam.

NOTES

262 FAO, 2016 (see note 138).

263 R.W. Kates. 2000. Cautionary tales: adaptation and the global poor. *Climatic Change*, 45(1): 5–17. J. Paavola and W.N. Adger. 2006. Fair adaptation to climate change. *Ecological Economics*, 56(4): 594–609; W.N. Adger, S. Agrawala, M.M.Q. Mirza, C. Conde, K. O'Brien, J. Pulhin, R. Pulwarty, B. Smit and K. Takahashi. 2007. Assessment of adaptation practices, options, constraints and capacity. In IPCC. 2007. *Climate change 2007, impacts, adaptation and vulnerability. Contribution of Working Group II to the Fourth Assessment Report of the Intergovernmental Panel on Climate Change*, pp. 717–743 [M. L. Parry, O. F. Canziani, J. P. Palutikof, P. J. van der Linden and C. E. Hanson, eds]. Cambridge, UK, Cambridge University Press; O.D. Cordona, M.K. van Aalst, J. Birkmann, M. Fordham, G. McGregor, R. Perez, R.S. Pulwarty, E.L.F. Schipper, and B.T. Sinh. 2012. Determinants of risk: exposure and vulnerability. In C. Field, V. Barros, T.F. Stocker, D. Qin, D.J. Dokken, K.L. Ebi, M.D. Mastrandrea, K.J. Mach, G.-K. Plattner, S.K. Allen, M. Tignor and P.M. Midgley, eds. *Managing the Risks of Extreme Events and Disasters to Advance Climate Change Adaptation.* A Special Report of Working Groups I and II of the Intergovernmental Panel on Climate Change. Cambridge, UK and New York, USA, Cambridge University Press.

264 FAO. 2016. *Climate change and food security: risks and responses.* Rome.

265 M.R. Carter and T.J. Lybbert. 2012. Consumption versus asset smoothing: testing the implications of poverty trap theory in Burkina Faso. *Journal of Development Economics*, 99(2): 255–264; H. Kazianga and C. Udry. 2006. Consumption smoothing? Livestock, insurance and drought in rural Burkina Faso. *Journal of Development Economics*, 79(2): 413–446; J. McPeak. 2004. Contrasting income shocks with asset shocks: Livestock sales in northern Kenya. *Oxford Economic Papers*, 56(2): 263–284; T. Kurosaki and M. Fafchamps. 2002. Insurance market efficiency and crop choices in Pakistan. *Journal of Development Economics*, 67(2): 419–453.

266 IPCC, 2014 (see note 58).

267 IPCC, 2012. *Managing the Risks of Extreme Events and Disasters to Advance Climate Change Adaptation. A Special Report of Working Groups I and II of the Intergovernmental Panel on Climate Change.* Cambridge, UK, and New York, USA, Cambridge University Press.

268 C.B. Field, L.D. Mortsch,, M. Brklacich, D.L. Forbes, P. Kovacs, J.A. Patz, S.W. Running and M.J. Scott. 2007. North America. In IPCC. 2007. *Climate change 2007, impacts, adaptation and vulnerability. Contribution of Working Group II to the Fourth Assessment Report of the Intergovernmental Panel on Climate Change*, pp. 717–743 [M. L. Parry, O. F. Canziani, J. P. Palutikof, P. J. van der Linden and C. E. Hanson, eds]. Cambridge, UK, Cambridge University Press.

269 Boko et al., 2007 (see note 79).

270 S.E. Eriksen and K.L. O'Brien. 2007. Vulnerability, poverty and the need for sustainable adaptation measures. *Climate Policy*, 7(4): 337–352; J. Ayers and S. Huq. 2009. Supporting adaptation through development: What role for ODA? *Development Policy Review*, 27(6): 675–692; E. Boyd and S. Juhola. 2009. Stepping up to the climate change: Opportunities in re- conceptualising development futures. *Journal of International Development*, 21: 792–804; J. Barnett and S. O'Neill. 2010. Maladaptation. *Global Environmental Change*, 20: 211–213; K. O'Brien, A.L. St Clair and B. Kristoffersen. 2010. *Climate Change, Ethics and Human Security.* Cambridge, UK and New York, USA, Cambridge University Press; L. Petheram, K. Zander, B. Campbell, C. High and N. Stacey. 2010. 'Strange changes': Indigenous perspectives of climate change and adaptation in NE Arnhem Land (Australia). *Global Environmental Change*, 20: 681–692.

271 J. Fanzo, R. McLaren, C. Davis and J. Choufani. 2017. *Climate change and variability. What are the risks for nutrition, diets, and food systems?* IFPRI Discussion Paper 01645. Washington, DC, IFPRI.

272 FAO, IFAD and WFP. 2015. *Strengthening resilience for food security and nutrition: a conceptual framework for collaboration and partnership among the Rome-based Agencies.* Rome.

273 Overseas Development Institute (ODI). 2016. *Resilience across the post-2015 frameworks: towards coherence?* London.

274 Established at COP 21 in Paris, France in 2015.

275 Established at COP 17 in Durban, South Africa in 2011.

276 Established at COP 7 in Marrakech, Morocco in 2001.

277 United Nations International Strategy for Disaster Reduction (UNISDR). 2017. Terminology. In *UNISDR* [online]. Geneva, Switzerland. https://www.unisdr.org/we/inform/terminology#letter-d

278 UNGA. 2015. Transforming our world: the 2030 Agenda for Sustainable Development. A/70/L.1. (21 October 2015).

279 ODI, 2016 (see note 273).

280 The Grand Bargain is an agreement between more than 30 of the world's biggest donors and aid providers, which aims to deliver an extra billion dollars over five years for people in need of humanitarian aid.

281 ODI, 2016 (see note 273).

282 UNSCN. 2017. *Sustainable diets for healthy people and a healthy planet.* United Nations System Standing Committee on Nutrition discussion paper. Rome.

283 R. Strohmaier, J. Rioux, A. Seggel, A. Meybeck, M. Bernoux, M. Salvatore, J. Miranda and A. Agostini. 2016. *The agriculture sectors in the Intended Nationally Determined Contributions: Analysis.* Environment and Natural Resources Management Working Paper No. 62. Rome, FAO.

284 WHO. 2016. Health and climate change – Report by the Secretariat. EB139/6. (20 May).

285 R.J.T. Klein, G.F. Midgley, B.L. Preston, M. Alam, F.G.H. Berkhout, K. Dow and M.R. Shaw. 2014. Adaptation opportunities, constraints, and limits. In IPCC, 2014 (see note 255), pp. 899–943.

286 Internal Displacement Monitoring Centre (IDMC). 2015. *Annual Report 2015.* Geneva, Switzerland.

287 R.W. Kates, W.R. Travis and T.J. Wilbanks. 2012. Transformational adaptation when incremental adaptations to climate change are insufficient. *Proceedings of the National Academy of Sciences,* 109(19): 7156–7161.

288 S.J. Vermeulen, B.M. Campbell and J.S.I. Ingram. 2012. Climate change and food systems. *Annual Review of Environment and Resources,* 37(1): 195–222.

289 UNSCN, 2017 (see note 282).

290 FAO, 2018 (see note 119).

291 UN. 2018. *UN Climate Resilience Initiative A2R* [online]. New York, USA. www.a2rinitiative.org

292 HLPE. 2012. *Food security and climate change. A report by the High Level Panel of Experts on Food Security and Nutrition of the Committee on World Food Security.* Rome.

293 M.V. Sánchez. 2018. Climate Impact Assessments with a Lens on Inequality.; UNSCN. 2016. *Impact Assessment of Policies to support Healthy Food Environments and Healthy Diets - Implementing the Framework for Action of the Second International Conference on Nutrition.* United Nations System Standing Committee on Nutrition discussion paper. Rome; UN. 2016. *World Economic and Social Survey 2016 - Climate Change Resilience: An Opportunity for Reducing Inequalities.* New York, USA.

294 WFP and Ministry of Economic Development of Sri Lanka. 2014. *Sri Lanka: Consolidated Livelihood Exercise for Analysing Resilience. A special report prepared by the World Food Programme and the Ministry of Economic Development.*

295 WFP. 2017. *How Climate Drives Hunger: Food Security Climate Analyses, Methodologies and Lessons, 2010–2016.* Rome.

296 FAO, 2016 (see note 138).

297 S. Asfaw, A. Scognamillo, G. Di Caprera, A. Ignaciuk and N. Sitko (forthcoming). *Rural livelihood diversification and household welfare: Cross-country evidence from sub-Saharan Africa heterogeneous impact of livelihood diversification.* Rome, FAO.

298 A.E. Boardman, D.H. Greenberg, A.R. Vining and D.L. Weimer. 2014. *Cost-Benefit Analysis: Concepts and Practice. 4th Edition.* The Pearson Series in Economics. Cambridge, UK, Cambridge University Press.

299 Least Developed Countries Expert Group. 2012. *National Adaptation Plans. Technical guidelines for the national adaptation plan process.* Bonn, Germany, UNFCCC Secretariat.

300 FAO. 2018. Integrating Agriculture into National Adaptation Plans (NAP-Ag) [online]. Rome. www.fao.org/in-action/naps

301 FAO. 2014. *The State of Food and Agriculture 2014. Innovation in family farming*. Rome.

302 WFP. 2017. *Climate Services*. Rome; WFP. 2018. *Climate Services* [online]. Rome. www1.wfp.org/climate-services

303 L. Lipper, N. McCarthy, D. Zilberman, S. Asfaw and G. Branca, eds. 2018. *Climate Smart Agriculture: Building Resilience to Climate Change*. Natural Resource Management and Policy. Berlin, Springer.

304 FAO. 2016. *Managing Climate Risk Using Climate Smart Agriculture*. Rome.

305 WFP and ODI. 2017. *Water for Food Security – Lessons learned from a review of water-related interventions*. Rome.

306 F. Baumhard, R. Lasage, P. Suarez and C. Chadza. 2009. *Farmers become filmmakers: climate change adaptation in Malawi*. Participatory Learning and Action. London, International Institute for Environment and Development (IIED). An example of a participatory video is the one conducted by IFRC in August 2008 in Salima, Malawi, with a focus on community-based adaptation to climate change; see IFRC. 2009. *Malawi: Adaptation to Climate Change by Mphunga villagers* [video]. www.youtube.com/watch?v=BwG1cW99ObM _

307 FAO. 2017. *Migration, Agriculture and Climate Change. Reducing vulnerabilities and enhancing resilience*. Rome.

308 M.V. Sánchez, 2018 (see note 293).

309 WFP. 2017. *Engaging stakeholders and building ownership for climate adaptation: best practice from Egypt*. Rome.

310 FAO. 2011. *The State of Food and Agriculture 2010–11 – Women in Agriculture: Closing the gender gap for development*. Rome.

311 WFP. 2018. *The R4 Rural Resilience Initiative* [online]. Rome. www1.wfp.org/r4-rural-resilience-initiative

312 WFP and OXFAM. 2016. *Impact Evaluation of the R4 Rural Resilience Initiative in Senegal, Final evaluation*. Rome; WFP. 2014. *HARITA / R4 Rural Resilience Initiative in Ethiopia, Impact Evaluation* [online]. Rome. www.wfp.org/content/harita-r4-impact-evaluation?_ga=2.6418226.1281503868.1516367150-1488310316.1490358925

313 World Bank. 2012. *World Development Report 2012. Gender equality and development*. Washington, DC, World Bank.

314 F. Branca, E. Piwoz, W. Schultink and L.M. Sullivan. 2015. Nutrition and health in women, children and adolescent girls. *British Medical Journal*, 351(h4173); I. Danton-Hill, C. Nishida and W.P.T. James. 2004. A life course approach to diet, nutrition and the prevention of chronic diseases. *Public Health Nutrition*, 7(1A): 101–21.

315 H. Alderman. 2010. Safety nets can help address the risks to nutrition from increasing climate variability. *The Journal of Nutrition*, 140(1): 148S–152S.

316 Inter-Agency Standing Committee (IASC). 2012. *Key humanitarian indicators*. Geneva, Switzerland; IASC. 2015. *IASC Emergency Response Preparedness Guidelines - July 2015 - Draft for field testing*. Geneva, Switzerland.

317 IASC, 2012 (see note 316).

318 GloPan. 2015. *Climate-Smart Food Systems for Enhanced Nutrition*. Policy Brief No 2. London, UK, Global Panel.

319 WFP. 2016. *Submission by WFP to the Subsidiary Body for Scientific and Technological Advice (SBSTA) on recent work in the area of climate impacts on human health*. Rome. WFP. 2018. *Submission by WFP to the Executive Committee of the Warsaw International Mechanism for Loss and Damage associated with Climate Change Impacts of the UNFCCC*. Rome.

320 WHO. 2015. *Operational framework for building climate resilient health systems*. Geneva, Switzerland.

321 WHO. 2013. *Essential Nutrition Actions - Improving maternal, newborn, infant and young child health and nutrition*. Geneva, Switzerland.

322 E. Wilkinson, L. Weingartner, R. Choularton, M. Bailey, M. Todd, D. Kniveton and C. Cabot Venton. 2018. *Forecasting hazards, averting disasters: Implementing forecast-based early action at scale*. London, ODI.

323 WFP. 2018. Food Security Climate Resilience (FoodSECuRE). In: *WFP Climate Change* [online]. Rome. www.wfp.org/climate-change/initiatives/foodsecure

324 S. Chantarat, C. Barrett, A.G. Mude and C.G. Turvey. 2007. Using weather index insurance to improve drought response for famine prevention. *American Journal of Agricultural Economics*, 89(5): 1262–1268.

325 J. Kellett and A. Caravani. 2013. *Financing risk reduction. A 20-year story of international aid.* London, ODI and Washington, DC, Global Facility for Disaster Reduction and Recovery.

326 C.P. Del Ninno, A. Dorosh and L.C. Smith. 2003. Public policy, markets and household coping strategies in Bangladesh: avoiding a food security crisis following the 1998 floods. *World Development*, 31(7): 1221–1238.

327 C.P. Del Ninno and M. Lundberg. 2005. Treading water: The long-term impact of the 1998 flood on nutrition in Bangladesh. *Economics and Human Biology*, 3(1): 67–96.

328 UNFCCC. 2007. *Investment and Financial Flows to Address Climate Change.* Bonn, Germany.

329 FAO. 2017. *Strategic work of FAO to increase the resilience of livelihoods.* Rome.

330 FAO, 2018 (see note 119).

331 WFP. 2016. *WFP Zimbabwe Situation Report #8.* [online]. Harare. https://documents.wfp.org/stellent/groups/Public/documents/ep/WFP284601.pdf

332 FAO, 2018 (see note 119).

333 IPC. 2017. *IPC Global Initiative 2017.* IPC Global Brief Series 2017. Rome.

334 UNGA. 2016. *Report of the open-ended intergovernmental expert working group on indicators and terminology relating to disaster risk reduction.* (1 December 2016); UNGA. 2017. *Implementation of the Sendai Framework for Disaster Risk Reduction 2015–2030.* (31 July 2017). NOTE: This report contains the updated and endorsed terminology related to disaster risk reduction contained in the note by the Secretary-General transmitting the report of the open-ended intergovernmental expert working on indicators and terminology related to disaster risk reduction (A/71/644). The agreed terminology can facilitate the implementation of the Sendai Framework and foster cooperation across and within nations, sectors and stakeholder groups. The agreement on terminology also helps countries and organizations to build shared understanding and foster coherent policy across the disaster risk reduction, sustainable development and climate change agendas (C. 28. A/72/259). It is coherent with the work of the Inter-Agency and Expert Group on Sustainable Development Goal Indicators, and the update of the publication entitled "2009 UNISDR Terminology on Disaster Risk Reduction" (see note 376).

335 FAO. 2013. *Resilient livelihoods: disaster risk reduction for food and nutrition security – 2013 edition.* Rome.

336 UNICEF. 2016. *Preparedness for emergency response in UNICEF - Guidance note.* New York, USA.

337 FAO, 2017 (see note 329). WHO, UNISDR and Public Health England. 2017. *Health Emergency and Disaster Risk Management Overview* [online]. www.who.int/hac/techguidance/preparedness/who-factsheet-overview-december2017.pdf

338 FAO. 2015. *Executive Brief: Tropical Cyclone Pam, Vanuatu.* Rome.

339 S. Thilsted, A. Thorne-Lyman, P. Webb, J.R. Bogard, R. Subasinghe, M.J. Phillips and E.H. Allison. 2016. Sustaining healthy diets: The role of capture fisheries and aquaculture for improving nutrition in the post-2015 era. *Food Policy*, 61: 126–131.

340 FAO, 2017 (see note 329).

341 FAO. 2015. *Nutrition and social protection.* Rome.

342 European Union (EU). 2012. *Les transferts sociaux dans la lutte contre la faim - Un instrument de référence pour les praticiens du développement Résumée.* Brussels and Luxembourg.

343 C. Cabot Venton. 2018. *Economics of Resilience to Drought in Ethiopia, Kenya and Somalia.* Washington, DC, USAID.

344 WFP. 2016. *Impact evaluation of the WFP Enhancing Resilience to Natural Disasters and the Effects of Climate Change programme with a specific focus on the resilience dimension.* Rome.

345 FAO, 2018 (see note 119).

346 UNGA, 2016 (see note 334).

347 L. Schäfer and E. Waters. 2016. *Climate risk insurance for the poor and vulnerable: How to effectively implement the pro-poor focus of InsuResilience*. Bonn, Germany, Munich Climate Insurance Initiative.

348 German Red Cross. 2017. *Forecast-based financing, an innovative approach*. Berlin.

349 FAO, 2013 (see note 335).

350 A. Savage, personal communication, 2018.

351 For a detailed description of the method, see: FAO. 2014. *Refinements to the FAO Methodology for Estimating the Prevalence of Undernourishment Indicator*. FAO Statistics Division Working Paper Series. Rome.

352 A person is considered healthy if his or her BMI indicates neither underweight nor overweight. Human Energy Requirements norms per kilogram of body mass are given in UNU, WHO and FAO. 2004. *Human energy requirements. Report of a Joint FAO/WHO/UNU Expert Consultation. Rome, 17–24 October 2001*. Rome.

353 See UN DESA. 2017. *World Population Prospects 2017* [online]. New York, USA. https://esa.un.org/unpd/wpp

354 See N. Wanner, C. Cafiero, N. Troubat and P. Conforti. 2014. *Refinements to the FAO methodology for estimating the prevalence of undernourishment indicator*. FAO ESS Working Paper Series ESS/14-05. Rome, FAO.

355 The EST division has developed and maintained a commodity balance database (XCBS) that provides elementary information for analysis of the food situation of a country or a group of countries. The XCBS provides balance sheet structured data for the major commodities in the following groups: cereals, dairy, meat, oil-bearing crops, sugar, tropical beverages, bananas and citrus. The data from the XCBS is used in a number of FAO publications and associated databases such as the Global Information and Early Warning System (GIEWS), Food outlook and crop prospects and food situation. XCBS provides up-to-date information on the state of agricultural commodity markets.

356 American Meteorological Society. 2015. *Glossary of Meteorology* [online]. Boston, USA. http://glossary.ametsoc.org/wiki/Weather

357 IPCC. 2012. *Managing the Risks of Extreme Events and Disasters to Advance Climate Change Adaptation*. A Special Report of Working Groups I and II of the Intergovernmental Panel on Climate Change. Cambridge, UK, and New York, USA, Cambridge University Press.

358 W.J. Cleveland and S.J. Devlin. 1988. Locally Weighted Regression: An Approach to Regression Analysis by Local Fitting. *Journal of the American Statistical Association*, 83(403): 596-610.

359 J. Bai and P. Perron. 1998. Estimating and Testing Linear Models with Multiple Structural Changes. *Econometrica*, 66: 47–78.

360 Integrated Food Security Phase Classification (IPC) (forthcoming). *IPC Technical Manual 3.0*.

361 As per Rome-based Agencies approach on Resilience: FAO, IFAD and WFP. 2015. *Strengthening resilience for food security and nutrition - A Conceptual Framework for Collaboration and Partnership among the Rome-based Agencies*. Rome.

362 As per UN Climate Resilience Initiative: UN. *UN Climate Resilience Initiative A2R* [online]. www.a2rinitiative.org

363 J. Agard, E.L.F. Schipper, J. Birkmann, M. Campos, C. Dubeux, Y. Nojiri, L. Olsson, B. Osman-Elasha, M. Pelling, M.J. Prather, M.G. Rivera-Ferre, O.C. Ruppel, A. Sallenger, K.R. Smith, A.L. St. Clair, K.J. Mach, M.D. Mastrandrea and T.E. Bilir, eds. 2014. Annex II: Glossary. In IPCC, 2014 (see note 66), pp. 1757–1776.

364 Agard *et al.*, 2014 (see note 363).

365 FAO. 2013. *Climate-Smart Agriculture. Sourcebook*. Rome.

366 As per Rome-based Agencies approach on Resilience: FAO, IFAD and WFP, 2015 (see note 361).

367 Agard *et al.*, 2014 (see note 363).

368 UNGA, 2016 (see note 334).

369 IPC (forthcoming) (see note 360).

370 Agard *et al.*, 2014 (see note 363).

371 Agard *et al.*, 2014 (see note 363).

372 IPCC, 2012 (see note 357).

373 Agard *et al.*, 2014 (see note 363).

374 Agard *et al.*, 2014 (see note 363).

375 R. Chambers and G.R. Conway. 1992. *Sustainable Rural Livelihoods: Practical Concepts for the 21st Century.* IDS Discussion Paper 296. Brighton, UK, IDS; S. Dercon, J. Hoddinott and T. Woldehanna. 2005. Shocks and consumption in 15 Ethiopian villages, 1999-2004. *Journal of African Economies,* 14(4): 559–585; WFP. 2009. *Comprehensive Food Security & Vulnerability Analysis (CFSVA) Guidelines - First Edition, 2009.* Rome; FAO, 2016 (see note 304).

376 Agard *et al.*, 2014 (see note 363). This glossary entry builds on the definition used in United Nations International Strategy for Disaster Reduction (UNISDR). 2009. *UNISDR Terminology on Disaster Risk Reduction.* Geneva, Switzerland, UN; and IPCC, 2012 (see note 357).

377 UNGA, 2016 (see note 334).

378 UNGA, 2016 (see note 334).

379 Agard *et al.*, 2014 (see note 363).

380 Agard *et al.*, 2014 (see note 363). This glossary entry builds on the definition used in UNISDR, 2009 and IPCC, 2012 (see note 357).

381 Agard *et al.*, 2014 (see note 363).

382 Agard *et al.*, 2014 (see note 363).

383 Agard *et al.*, 2014 (see note 363).

384 UNGA, 2016 (see note 334).

385 Agard *et al.*, 2014 (see note 363).

386 Agard *et al.*, 2014 (see note 363).

387 UNGA, 2016 (see note 334).

388 UNGA, 2016 (see note 334).

389 UNGA, 2016 (see note 334).

390 UN Chief Executives Board for Coordination (CEB). 2017. *Report of the High-Level Committee on Programmes at its thirty-fourth session.* Annex III. CEB/2017/6 (6 November 2017).

391 As per Rome-based Agencies approach on Resilience: FAO, IFAD and WFP, 2015 (see note 361).

392 UNGA, 2016 (see note 334).

393 American Meteorological Society, 2015 (see note 356).